ALASKA'S
Copper River Delta

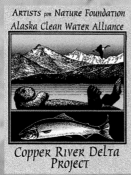

Artists for Nature Foundation
Alaska Clean Water Alliance

Copper River Delta
Project

Published by
ARTISTS FOR NATURE FOUNDATION
Netherlands
in association with
THE UNIVERSITY OF WASHINGTON PRESS
Seattle

THE UNIVERSITY OF WASHINGTON PRESS

ALASKA'S

Copper River Delta

Riki Ott

Foreword by Paul R. Ehrlich and
Anne H. Ehrlich

TO MY FATHER AND MOTHER, FRED AND JOLLY OTT,
AND TO THE PEOPLE OF THE COPPER RIVER –
PAST, PRESENT, AND FUTURE.
(RIKI OTT)

Library of Congress Cataloging–in–Publication Data
Ott, Riki.
Alaska's Copper River Delta/Riki Ott.
160 p., 25.4 x 19.5 cm.
Includes bibliographical references (p. 152) and index.
ISBN 0–295–97743–4 (alk. paper).
1. Natural history – Alaska – Copper River Delta. 2. Nature conservation – Alaska –
Copper River Delta. 3. Wetland ecology – Alaska – Copper River Delta. 4. Natural his-
tory – Alaska – Copper River Delta – Pictorial works. 5. Nature conservation – Alaska
– Copper River Delta – Pictorial works. 6. Wetland ecology – Alaska – Copper River
Delta – Pictorial works. I Artists for Nature Foundation. II Alaska's Copper River Delta.
QH105.A4088 1998
508.798'3 – dc21
98–25746
CIP

ISBN 0–295–97743–4 (alk. paper).

The paper used in this publication meets the minimum requirements of American
National Standard for Information Sciences–Permanence of Paper for Printed Library
Materials, ANSI Z39.48-1984.

ILLUSTRATIONS

FRONT COVER
Vadim Gorbatov
*Bald Eagle over the Copper
River Delta*
Watercolor, 54 x 57 cm

BACK COVER
Victor Bakhtin
*Bald Eagle, Childs Glacier
and Copper River*
Watercolor, 18 x 32 cm

TITLE
Colin See-Paynton
Great Horned Owl
Wood engraving, 34 x 16 cm

TITLE PAGE
David Barker
Ridge
Watercolor, 37 x 55 cm

CONTENTS
Victor Bakhtin
Hook Point
Gouache, 29 x 62.5 cm

CONTENTS

Art is a universal language. In this book the Artists for Nature Foundation uses the unique vision of artists from around the world to bring attention to Alaska's Copper River Delta. The Delta is one of America's most remarkable wetlands, a place superlative even by Alaska's standards. It is an enormously productive ecosystem and critical habitat for countless migratory birds, salmon and other wildlife. The Delta is also next-door neighbor to Prince William Sound, site of the 1989 Exxon Valdez oil spill. Thus the Delta provides an important refuge for many wildlife species still recovering from that environmental disaster.

In the wake of the oil spill, residents of the region are asking important questions about how to protect their commercial and subsistence fisheries and other natural resource-based economies -- so that they may be healthy for generations to come and not sacrificed for short term profit. These people are seeking ways to work together to protect their communities and special way of life based on respectful sharing of the Copper River Delta's natural bounty. I salute their efforts.

As former President and as private citizen, I have always held Alaska dear to my heart. I hope all readers of this book will be moved as I am by the vision of these artists and local communities in celebrating the beauty of the Copper River Delta.

Jimmy Carter

< [1] **Childs Glacier calving**

Andrew Haslen: "The glacier is breathtaking. I stand across the half-mile-wide Copper River watching blocks of ice the size of apartment buildings fall into the water, first with a huge explosion of water, then the sound of thunder and the tidal wave crossing the river. Feeling small and vulnerable, I put my sketchbooks away and just watch. I have no idea how to capture it on paper, and no intention of even trying."

(PHOTO: PAT AND ROSEMARIE KEOUGH)

DELTAS - COPPER AND OTHERWISE

One of the biggest thrills of our lives was flying in a light plane over the Copper River Delta in May of 1994. Even though migration was not at its maximum, the concentrations of shorebirds - especially western sandpipers, dunlins, and other 'peeps' - were breathtaking, one of the great wildlife scenes on planet Earth. It was a birdwatcher's paradise, but many other thoughts ran through our minds. Paul was overwhelmed with nostalgia, with thoughts of times more than 40 years earlier when he studied butterflies and watched birds in Canada's arctic and in Alaska. There's a feeling in the North that's now hard to find on most of the rest of Earth, a feeling that human beings have not yet completely dominated the landscape. It's a feeling that many of us would love our grandchildren to be able to experience, and we realize that only dramatic efforts have any chance of saving the few remaining places where it is possible.

And this still unspoiled region is one of the world's loveliest: backdropped by spectacular coastal mountains, and containing a vast temperate rain forest as well as the lush wetlands of the delta itself.

Cued by the swarms of birds, wheeling in well-coordinated flocks beneath our plane, both of us thought of the very special functions deltas often perform in the ecology of our planet. Flat places with braided streams where fresh and salt waters mix are often rich centers of diversity, and that diversity can play critical roles in providing ecosystem services – such as supplying humanity as well as shorebirds with food from the sea. Indeed, deltas were among the first places where human beings found resources sufficiently abundant that they could give up their nomadic ways and settle down in one place. In the northwestern part of North America, sedentary groups of Native Americans depended upon the rich annual harvest of salmon that could be obtained in river deltas, and also gained nourishment from many other marine organisms and some forest plants and animals. And, as is traditional in human societies, they fought wars over those resources before the dawn of history. The rich soils of river valleys and deltas have also provided humanity with some of its best farmland. One need only think of the Egyptians, who built their civilization on the riches of the Nile delta.

Thoughts about deltas can thus be quite cheering, but in other contexts they can be quite depressing. In intensely overpopulated Bangladesh, people are forced to farm the rich soils of the Ganges delta, often deposited in the form of 'chars' where the elevation is just a few feet above sea level. Periodically, great typhoons sweep up the Bay of Bengal, and at times they have caused catastrophic flooding in which hundreds of thousands of people swept off the chars have lost their lives. Deltas can be precarious places for Homo sapiens as well as generators of important nourishment. And the silt that builds deltas, if supplied in too much abundance by tropical rivers with deforested watersheds, can have lethal effects on the coral reefs lying just offshore.

Ecologically, the Copper River Delta provides us with one of the best examples of both biological abundance and ecological precariousness. The hordes of shorebirds that gather to feed there each spring on their way north have no other areas sufficiently productive to 'refuel' their vast numbers after exhausting migratory flights. If the Copper River Delta were to disappear, so too would the majority of shorebirds that migrate along our western coast, especially the western sandpipers and dunlins. What the long-term results of the decimation of those populations would be is difficult to say, but at the end of March 1989 humanity almost ran the experiment that would have provided the answer. The *Exxon Valdez* oil spill in nearby Prince William Sound nearly engulfed the Delta. The shorebird populations were spared by pure happenstance.

So as you go through this magnificent book, think not just of the beauty and fascination of the Copper River Delta region. Think also of the important roles that deltas play directly or indirectly in all of our lives. They are precious natural areas, increasingly subjected to ecological deterioration for which our descendents are likely to pay a very high price. The people of the Copper River Delta region and the town of Cordova are dedicated to preserving its values, and they still have some chance of developing a sustainable economy based on that preservation. We hope that this magnificent book will help encourage and support that effort.

Paul R. Ehrlich and Anne H. Ehrlich

Center for Conservation Biology
Department of Biological Sciences
Stanford University

ACKNOWLEDGMENTS

The author wishes to thank each and every person who contributed to this project. Your time, energy, good will, services, and financial support made a monumental undertaking possible and enjoyable. I humbly and gratefully acknowledge your trust and faith in my abilities to do the things I said I would do. As listing you all would have left little room for text and art, I am reduced to giving special thanks to the following individuals and organizations.

To Gershon Cohen and the past and present board and staff of the Alaska Clean Water Alliance for sponsoring and nurturing the Copper River Delta Project for three years until it was strong enough to take off on its own.

To Ysbrand Brouwers, Wil Bouwmeester, and past and present board of, and participating artists with, Artists for Nature Foundation for turning a dream into reality.

To Cordova Visual Artists for moral and logistical support with the visiting artists and art tours.

To Sylvia Lange and Greg Meyer, owners of Cannery Row, for hosting the visiting artists at Cannery Row, and for changing the working "Net Loft" into the "Art Loft," a studio with an inspirational atmosphere and history, and the fishermen, who found themselves mending their nets outside because of the change in circumstance, for their understanding and good humor.

To Gayle Ranney and Steve Ranney, with Fishing and Flying, for orientation flights and air charter drop-offs and pick-ups.

To Paul Kelly, manager of Davis Groceries, for taking the plunge and being the first Cordova business to contribute cold, hard cash to the project: you channeled the flood waters. And to all the other individuals, businesses, foundations, and public agencies that helped sponsor the project: every little bit helps.

Gracious thanks for timely funding to Alaska Conservation Foundation, the Alaska Department of Environmental Conservation, the Bullitt Foundation, the Charles Stewart Mott Foundation, the Leighty Foundation, the National Forest Foundation, Recreational Equipment, Inc., the Skaggs Foundation, the United States Environmental Protection Agency, and the Zoological Society of Milwaukee.

To ad hoc steering committee members Mike Anderson, Cal Baker, Max Bennett, Nancy Bird, Dr. Mary Anne Bishop, Becky Chapek, Belen Cook, Kim Ewers, Bob Henrichs, Mark Hoover, Scott Janke, Brian Lettich, Roy Nowlin, Dave O'Brien, Linden O'Toole, Liz Senear, Cheri Shaw, Kristin Smith, Karen and Paul Swartzbart, and Glenn Ujioka for your early perception of, and selfless support for, possibilities for change.

To founding board members Jack Hopkins, Wilson Justin, Sylvia Lange, and Dan Stevens, and staff Kristin Smith, Lisa Marie Van Dyck, and Chris Beck for your collective wisdom, good natures, visionary thinking, attention to detail, and commonsense aptitude for turning theory into practice: what a team!

To Tony Angell, artist and friend, for your unflagging faith, patience, quick thinking, and good humor in putting the right players together to make this book and art tour happen.

To all who read the manuscript, offering corrections and suggestions, including Steve Babler, Cal Baker, Chris Beck, Karl Becker, Dan Bilderback, Nancy Bird, Dr. Mary Anne Bishop, Dave Blanchet, George Covel, Brent Davies, Ruth Fairall, Sandy Frost, Steve Grabacki, Bob Henrichs, Barcley Jones, Wilson Justin, Sylvia Lange, Brian Lettich, Dan Logan, Ralph Lohse, Catherine Mater, Dr. Pete Mickelson, Roy Nowlin, Cathy Sherman, Kristin Smith, Linden O'Toole, Vic Van Ballenberghe, John Wilcock, and Don Yukey, and especially David Grimes for your meticulous and patient word-smithing, and to 11-year-old Makena O'Toole, young woodsman, for all your helpful suggestions in the ecology chapter.

To Gretchen Van Meter, my editor with the University of Washington Press, for your patient rearranging of the text into a focused story.

And especially to Larry Merculieff, mentor and friend, for your infinite good will, radiant energies, and positive thoughts: you helped me see the Endless Possibilities Within.

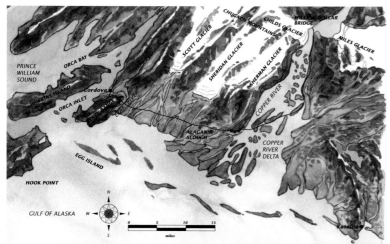

MAPS: SUSAN OGLE

The Copper River at a Glance

Fifth largest river system in Alaska

287 miles (465 km) long

Drains 26,500 square miles (70,000 km^2) or almost 17 million acres (7 million ha)

Headwaters in the highest coastal mountains in the world (9 mountains over 16,000 feet or 4,850 m)

Highest point in drainage: Mt. Logan 19,850 feet (6,015 m), second highest peak in North America

Drains tributaries of Bering River-Bagley Icefield complex, including world's largest glacier (twice the size of Rhode Island) outside of Greenland and the polar regions.

Estimated annual sediment load of 80 million metric tons, one-third that of the Mississippi River, which drains 40 times more area

Summer floods of 500,000 cubic feet (140,000 ds) per second

Estimated annual discharge of 38 million acre feet (4.5 million hectares to meter depth)

Largest contiguous wetland on the Pacific coast of North America, at 700,000 acres (283,000 ha) and 70 linear miles (113 km) wide

OVERVIEW

Water is life.
We are the people who live by the water.
Pray by these waters.
Travel by these waters.
We are related to those who live in the water.
To poison the waters is to show disrespect for creation.
To honor and protect the waters is our responsibility
as people of the land.

Winona LaDuke
Anishinabe Nation, Minnesota

COPPER RIVER COUNTRY

On a particular stretch of the remote North Gulf coast of Alaska, there is a place where the mountains, rivers, and sea work together to create an extraordinary wetlands. This 700,000 acre (283,000 hectares) wetlands, the largest on the Pacific coast of North America, is called the Copper River Delta.

It is actually a composite of contiguous deltas from all the waterways draining into the Gulf of Alaska between Cape St. Elias on the east to the Heney Range on the west, near the fishing village of Cordova, a stretch of 70-odd miles (113 km). The Delta is a dynamic landscape, constantly being reshaped by violent floods and earthquakes, by storms and currents in the Gulf of Alaska, and by the huge volume of silt ground from the mountains by glaciers and carried by the river to the sea.

The Delta is dominated by the glacial Copper River, draining, over its 287-mile length (465 km), the fifth largest watershed in Alaska. The Copper River begins in the interior of southcentral Alaska, in rolling boreal taiga forests of black spruce, white spruce, and birch, shadowed by spectacular peaks of the Alaska Range and Wrangell–St. Elias Mountains. This is a land of warm summers and deep, cold winters. Precipitation is low, about 11 inches (28 cm) a year. In this open, interior landscape, the Copper is joined by glacial tributaries such as the Tazlina, Tonsina, Klutina, Chitina, and Tasnuna (the suffix 'na' means water or river in Athabascan). The river then flows south, cuts mile-deep canyons through the snowy peaks of the coastal Chugach Mountains, squeezes between two glacier faces, and emerges in a very different world – the rain forests, rich estuaries, and saltwater smells of the Copper River Delta.

Even amidst the grandeur and wonder of Alaska, the Copper River watershed stands out. The mountains that cradle the Copper include nine of North America's fifteen tallest peaks, making this the highest coastal range in the world. These towering peaks block passage of intense low pressure systems generated in the Gulf of Alaska, producing tremendous amounts of precipitation. Captured by the high peaks, wet snow compacts over time into huge masses of ice, creating glaciers and icefields,

[2] **Victor Bakhtin**
Bald Eagle, Childs Glacier
and Copper River
Watercolor, 18 x 32 cm

[3] **Piet Klaasse**
Eva Sinyon
Conté, 28 x 19 cm
The Eyak culture today is represented by some 325 survivors in their last stronghold, the Copper River Delta.

[4] **Childs Glacier and**
Copper River
(PHOTO: PAT AND ROSEMARIE KEOUGH)

one of which, the Bering Glacier–Bagley Icefield complex just east of the Delta, is the largest ice mass outside of the polar regions and Greenland.

The 100 to 160 inches (254 to 406 cm) of rain that falls annually at sea level on the Delta also produces a lush rain forest. This is a rare, subarctic, coastal temperate rain forest, the northernmost extension of a forest that continues along the coast through southeast Alaska and British Columbia to northern California. Of the five Pacific Northwest states and provinces with coastal rain forests, Alaska has the largest proportion of undeveloped, mature rain forests, consisting mostly of Sitka spruce and western hemlock. Only eleven of forty-six watersheds larger than 250,000 acres (100,000 hectares) in the Pacific Northwest remain intact, and the Copper River Delta contains two of them.

The Copper River Delta is the crown jewel of North America's wetlands. The Delta is a key stopping area for the Pacific coast flyway during migratory periods: it is designated as a hemispheric site in the Western Hemisphere Shorebird Reserve network. The Delta is also designated as a Critical Fish and Wildlife Habitat by the State of Alaska. The Copper River and the myriad streams of the Delta are pristine spawning grounds for world-renowned

< [5] **David Rosenthal**
Delta from Heney Ridge
Oil, 71 x 76 cm
The Delta is the crown jewel of North America's wetlands. Here is a corner of the western Delta, looking east towards Sheridan and Sherman Glaciers.

[6] **David Rosenthal**
Little Mummy Island
Oil, 137 x 152 cm
Looking south from Cordova, the Copper River Delta is to the east and Prince William Sound to the west.

wild sockeye, silver, and king salmon. Moose, brown and black bear, wolves, endangered Steller sea lions, and a host of other mammals, fish, and birds use the Delta during the summer months or year round.

COMMUNITIES: ECONOMY AND LAND USE PATTERNS

There are 5,600 human residents in this 17 million-acre (7 million hectares) watershed, half of whom live in the seaport of Cordova, the only community on the Delta and the only incorporated town in the region. The other half live in the upper watershed (Copper Basin) in twenty-one small settlements mostly located on tributaries of the river. The upper watershed is on the state road system, but the lower watershed can be reached only by boat or plane. In general, year-round job opportunities in the region are limited. Average payroll earnings are lower than in other parts of the state, and the cost of living is over one-third higher than that of Anchorage (which is itself twenty percent higher than that of Seattle, Washington). Alaska Natives account for about twenty percent of the population, and well over ninety percent of the residents practice a subsistence lifestyle, harvesting and sharing natural resources. Subsistence activities are an important supplement to income for people in a region where over sixteen percent of the population lives below the poverty level.

Land ownership patterns dictate development. Over ninety percent of the watershed is public land. A large portion of the Wrangell–St. Elias National Park and Preserve makes up the northeastern portion of the watershed; at 13 million acres (5.3 million hectares), it is the largest national park in the United States, and, together with the contiguous Kluane National Park of Canada, comprises the largest protected parkland in the world. The Bureau of Land Management and the State of Alaska manage 2.3 million and 3.3 million acres (930,000 and 1.3 million hectares), respectively, in the Copper

Basin. The Delta is part of the Chugach National Forest, the nation's second largest forest system. Among all national forests, the Delta alone is managed by a special federal law with a priority for fish and wildlife habitat; in all others, multiple use including harvestable resources is the management priority.[1] Most of the accessible, more readily developable land in the region is private, including private inholdings within the public land boundaries. The large majority of these private lands is held by Native corporations.[2]

The rural economy is driven by harvest of natural resources (primarily fish and, recently, timber), although tourism is steadily increasing throughout the watershed. Currently there is no comprehensive long-term plan for resource development or management in this watershed, mainly because of

< [7] **Hartney Bay shorebirds**
Every spring the Delta experiences one of the world's greatest shorebird migrations when some 5 to 7 million, representing as many as 36 species, stop to feed on the Delta's superabundant food reserves. The vast majority are western sandpipers (pictured here) and dunlins.
(PHOTO: YSBRAND BROUWERS)

[8] **Sandpiper flock**
Ninety percent of the migrants along the Pacific coast flyway stop to rest and feed on the Delta.
(PHOTO: COURTESY OF U.S.F.S.)

[10] *Sheridan Glacier and Delta pond*

(PHOTO: DAVID GRIMES)

the political boundaries of different land owners. While fisheries resources have been managed by the State of Alaska for sustained yield, forestry resources have not. Tourism growth is haphazard, responding more to pressure by urban and industrial interests than to comprehensive planning by residents within the watershed. If the current lack of planning continues, natural resource capital within the watershed will be exhausted and local economies will collapse or will continue in a boom-bust style.[3]

Alaska has been called the 'take-out state' for its numerous extractive industries. The Copper River watershed is no exception: copper extraction by the Guggenheim-Morgan syndicate in the early 1900s, fish processing by non-Alaska-based companies, and logging by multinational timber companies have all generated large profits, but little of this wealth has stayed in the region. Dependence on low value-added commodity exports has made the regional economy particularly vulnerable to changes in markets and commodity prices.

< [9] *Delta ponds and sloughs*

(PHOTO: COURTESY OF U.S.F.S.)

THE *EXXON VALDEZ* OIL SPILL AND THE COPPER RIVER WATERSHED PROJECT

Turning points often revolve around one central, catalyzing event.

On Good Friday, March 24, 1989, the *Exxon Valdez* spilled oil into Prince William Sound adjoining the Delta to the west, damaging many of the area's fisheries and subsistence resources. Nine years later, scientists reported that only one species, the bald eagle, had fully recovered from the spill.[4] Lingering effects and delayed recovery were found in many species, including the Pacific herring, a species absolutely critical to the food web in the sound. The socio-economic effect on Cordova during and since the spill has been devastating.[5]

To make matters worse, residents were besieged with a flood of large-scale, unsustainable development plans, many of which are still pending or were resurrected in another version after the initial plan was defeated. These plans included oil and gas lease sales, proposed construction of a major highway across the Delta and a deep-water port near Cordova for resource extraction and large-volume (cruise ship) tourism, and the proposed clear-cutting of private lands (within the Chugach

National Forest). When clear-cutting and highway construction began, residents found themselves reacting to the plans and actions of others instead of actively working together to initiate a local, comprehensive planning process. There's a saying in Cordova that if something is important, half the town will be for it and half against it. The town was rocked by lawsuits, petitions, and protests as issues of economic diversification polarized the residents.

Local government held a series of town meetings in 1994-95 to determine the community's desires. Residents agreed that efforts to diversify the economy should also maintain the quality of life and protect the Copper River ecosystem (the town's only fishery undamaged by oil). But how to get 'there' from 'here'?

It started when two dozen individuals representing the diverse interests within the community came together and agreed to listen to one another. Gradually, the members of this group learned that their greatest obstacle was one of their own making: a lack of mutual trust. By working together, trust grew. So did enthusiasm for problem-solving and hope for the future. As Ralph Waldo Emerson once wrote: "Every great and commanding movement in the annals of the world is the triumph of enthusiasm."

Working together as the Copper River Delta Project, this representative group of citizens decided to take action to foster economic development in ways that allowed for long-term, sustainable resource use and benefits to residents of the watershed. The group decided to deal with short-sighted economic interests by offering better ways of doing business for public and private land owners, rather than by confrontation; for example, certified sustainable forestry instead of clear-cutting, and ecotourism with niche-marketing instead of large-scale cruise ship or road traffic with commodities-marketing.

The group slowly forged a dedicated, broadly

< [11] **Bruce Pearson**
Egg Island and Beyond
Watercolor, 68 x 12 cm
This barrier island hosts the largest glaucous-winged gull colony in the world, with some 10,000 breeding pairs.

[12] ***Bruce Pearson painting***
(PHOTO: YSBRAND BROUWERS)

collaborative alliance of business, Native, and environmental interests, resource agencies, scientists, and individuals working together for sustainable development. Over time, as residents throughout the far-flung reaches of the community became interested in linking their efforts to create more diverse and sustainable communities throughout the region, the Copper River Delta Project evolved into the Watershed Project. By linking their efforts, residents are finding they have a stronger voice in decisions affecting their future. The Copper River Watershed Project is an on-going process of collaborative planning, decision-making, and action.

ARTISTS FOR NATURE FOUNDATION

One of the Project's earliest partnerships was with the Artists for Nature Foundation, a group of kindred spirits based in the Netherlands. This group of over one hundred artists from more than twenty different countries is dedicated to promot-

[13] **David Barker**

Midnight Sun (triptych)

Pastel / acrylic, 45 x 54 cm (x3)

The Delta is a dynamic landscape shaped by earthquakes, storms, floods and the huge volume of sediment carved from the mountains by glaciers and carried by the Copper River to the sea. "Here is a view of the Delta based on my very first impressions of Alaska as our flight dropped below the clouds shortly before midnight. Our responses to our new surroundings are as curious as curiosity itself. From microscopic to panoramic, each artist selects his or her special interest."

then used to draw worldwide attention to the value of these at-risk areas and, ultimately, to protect them through alternative resource use and conservation strategies.[6]

A SENSE OF PLACE

This book, then, is a story about a sense of place and about the search for a sense of the value of that place in the great Copper River Delta. Chapter 2 explores the formation of the Delta and the ongoing geologic, tectonic, and climatic physical forces at work in this region; chapter 3 adds the living layer by exploring the ecosystem from the ocean through the wetlands and forests to the alpine tundra and rocky slopes; and chapter 4 reviews human activities in this region, from ten thousand years ago to the present. Appendixes A and B discuss the *Exxon Valdez* oil spill and the Copper River Watershed Project, respectively, presenting choices for a sustainable future for the communities in the region.

The art portrayed in this book hints at what awaits discerning visitors who make the extra effort to get to Cordova and take the extra time to stay and explore the community and the Copper River Delta. By keeping this quiet community a little off the beaten track, it remains a pocket of the real Alaska, an unspoiled coastal fishing community with a rich history and stunning surroundings.

ing, through the unique perspective of artists, conservation of natural and historical landscapes as an essential element of sustainable community development. The Foundation serves as a catalyst between the public and nature conservation organizations, using the universal language of art to achieve its goals in a positive way. The organization focuses on little-known areas with globally significant resources facing large-scale development or other threats. The Foundation organizes teams of accomplished artists from around the world and brings them into the threatened regions. Once on site, the artists utilize their vision and talents to portray the interrelationships between the land, its wildlife, and its human inhabitants. The artwork and other products resulting from the program are

[15] **Andrea Rich**
Ravens in the Hemlock
Woodcut, 33 x 42 cm
Ravens have a large vocabulary of calls, which may include names for each other. In coastal Native mythology, Raven is a trickster and creator. "On a cool wet evening the mist was surrounding the hemlock trees, and it looked like a scene from a fairy tale complete with characters."

< [14] **Bruce Pearson**
Observation Island
Mixed media, 70 x 115 cm
The Delta's forests are in the northernmost part of a coastal temperate rain forest that runs in an arc 2,000 miles (3,000 km) all the way from Kodiak to California.

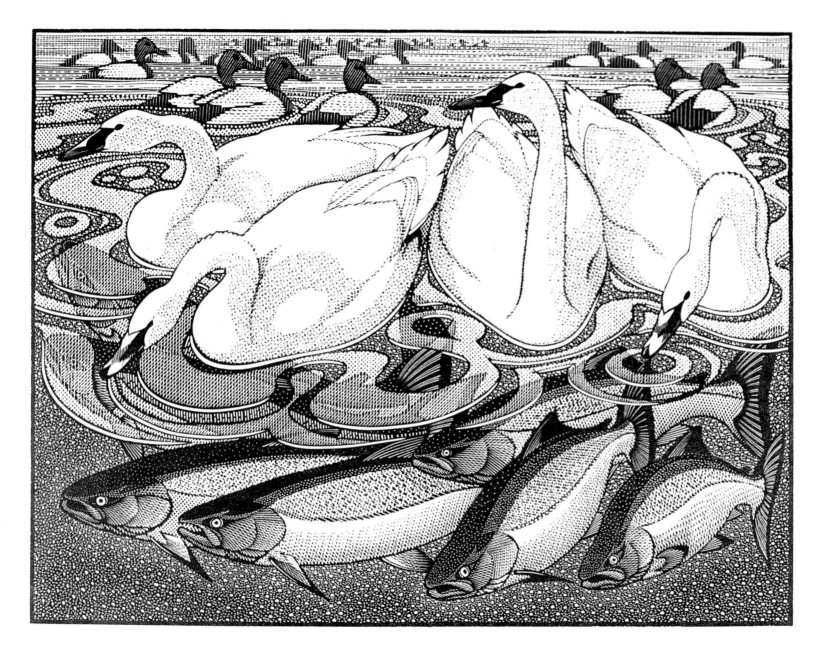

'Sandpiper Party'

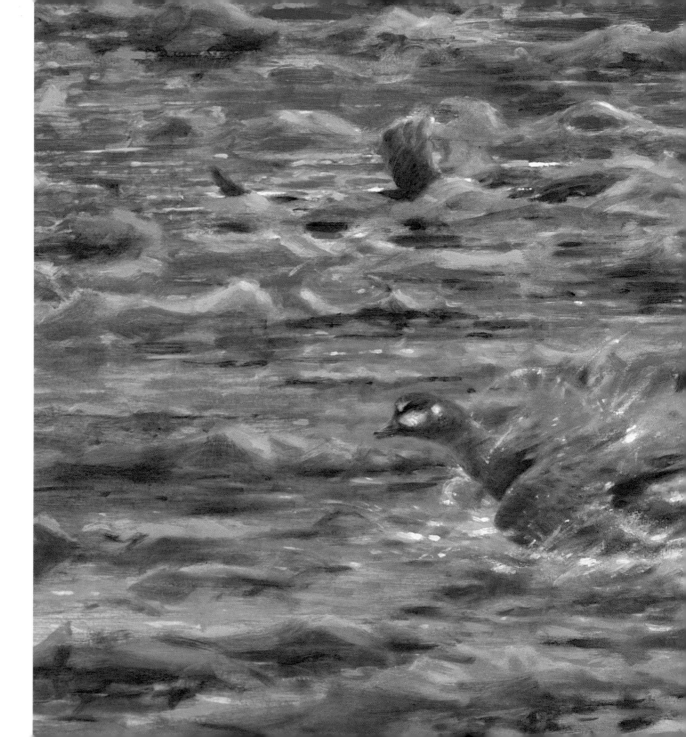

[18] **Bruce Pearson**
Harlequin Display
Oil on paper, 60 x 105 cm
*Sea otters and harlequin
ducks are among the numerous species yet to recover
from the 1989 Exxon Valdez
oil spill in Prince William
Sound, just west of the
Delta. Fortunately the Delta
was not touched by the spill
and remains a sanctuary for
the largest concentration of
sea otters on the Pacific coast.*

Delta movements II — Sheridan Glacier

[20] **Colin See-Paynton**
Delta Movements II
Wood engraving, 21.5 x 26 cm

[21] **David Bennett**
Steller Sea Lions and Black Oystercatchers
Mixed media 33.5 x 47.5 cm
The Delta hosts one of the west-ernmost healthy breeding pop-ulations of Steller sea lions in North America. In western Alaska, Steller sea lions are in severe decline. The species has been listed by the federal government as endangered.

[22] **David Bennett**
Beaver's Lodge Watch
Watercolor, 54 x 73 cm
Moose, the largest members of the deer family, frequent willow-
alder thickets, marshes, and shallow ponds in secluded areas of
the Delta. These moose are all descendants of twenty-one calves
flown in from other areas of Alaska between 1949–58 by legenda-
ry Cordova bush pilot 'Mudhole' Smith. Constant tinkering by one
of the world's largest concentrations of beaver reshapes the Del-
ta's drainages.

THE PEOPLE
  Formed a circle round the Fire,
    each showing an attentive face
      to every other person.

AND THEY SPOKE,
  each waiting quietly
    till the other had finished,
      as they had learned to do,
  a circle of silent listening
    framing the wisdom each contained
  until the wisdom of all was spoken,
    contained at last
      by the Circle of the People.

Thinking again
  of those earlier days
    when much noise
      but little wisdom was heard,

Thinking now
  of the quiet circle of listening hearts,
    they were filled with an understanding
      of the value of their way.

AND A FIRM RESOLVE SWEPT THROUGH THEM.

THEY DECIDED
  To be a People
    who would perpetuate and refine
      this manner of ordered council
        which they had achieved

So that the children's children's children
  might benefit from greater understanding . . .

And their paths through joy or sorrow
  might be eased
    by the soft sounds of wisdom's voice.

For they saw the People
  like a Great River –
    spreading out upon the land,
    spreading out across the waters,
      dividing down a thousand thousand paths
        not yet seen.

AND A SENSE OF TOMORROW
  ENTERED THEIR HEARTS
    AND NEVER AGAIN LEFT THEM.

SUCH WISDOM IS OUR GIFT
  FROM THOSE WHO WENT BEFORE.

MAY WE OFFER EQUAL MEASURE
  TO THOSE WHO FOLLOW US.

Paula Underwood
from The Walking People,
A Native American Oral History[7]

< [23] **David Barker**
Bears
Acrylic, 41 x 55 cm
*"Even large and individually
spectacular animals like
bears appear as small parts
of an infinite landscape from
an aerial perspective, one that
reveals the textures, patterns
and colors of the Delta."* These
brown bears of the Alaska
coast grow larger than their
interior cousin, the grizzly,
because of the plentiful sup-
ply of salmon. Red salmon,
or sockeye, pictured here
are in their distinctive
spawning color.

GENESIS & CLIMATOLOGY

FORMATION OF THE COPPER RIVER AND ITS DELTA

ANCIENT TECTONIC FORCES

From 50 to 5 million years ago, during the Tertiary period, the mountains that cradle the Copper River today were created when parts of Earth's crust, riding high atop shifting oceanic tectonic plates, collided with a continental plate to form southern Alaska. Prior to this collision, the land mass that rafted in to become southern Alaska had had a long history. Some of the oldest rocks on this drifting mass had formed in underwater volcanic eruptions in equatorial seas 240, 300, and even 500 million years earlier. Lumped together by violent collisions of ancient mid-ocean tectonic plates, the different rock terranes, each with its distinct history, then drifted as a single Superterrane for about 200 million years, moving slowly northeast on the Pacific Plate. This microcontinent first collided with the North American Plate near southern British Columbia about 110 million years ago. There it stuck for over 50 million years until oceanic plates, shifting and pivoting, tore it loose and drove it north again. For 10 million years this Superterrane lurched along the edge of the continental plate until, about 50 million years ago, it struck with such force that it once again stuck fast to become southern Alaska.

For the next 10 million years, the leading edge of the Superterrane bulldozed into the North American Plate, crumbling the ocean basin and continental shelf in great wavelike folds. A time-lapse film of this activity sped up would resemble an Eskimo blanket toss. As the Pacific Plate subducted, or dove under, the continental plate, intense pressure melted the wet sediments, water, and crust into a light magma, which rose to the surface, far inland, to become the Alaska Range. Small bits of the

[25] **David Rosenthal**
Wind-Bent Willow
Oil, 71 x 76 cm
The Copper River valley is a wind tunnel between the interior and the coast. Through time the silt-laced winds have built sand dunes on the upper Delta.

[26] *Glacier crevasses*

(PHOTO: PAT AND ROSEMARIE KEOUGH)

Superterrane tore off and continued to move in a counterclockwise direction along the coast in a 'strike-slip' fashion, much like a deck of cards smeared across a table top, forming the arcuate shape of the southern coastline of Alaska.

About 20 million years ago, a smaller land mass, following the track of the Superterrane, struck the

< [24] **David Rosenthal**
Full Moon, Sheridan Glacier
Oil, 71 x 76 cm
The headwaters of the Copper River lie within the highest coastal mountain range in the world, draining the largest glaciers outside of Greenland and the polar regions. This frozen, iceberg-studded lake at the terminus of the Sheridan Glacier is, in winter, a favorite skating area for locals.

southern coast of Alaska. For over 15 million years the leading edge of this latest addition thrust violently under the continental margin, uplifting the Wrangell–St. Elias and Chugach Mountains as the oceanic plate subducted into the Aleutian Trench off Kayak Island. Some uplift from this collision is still occurring.

The mountains born of Earth's labor rimmed a huge basin, bordered on the north by what now is known as the Alaska Range, on the east by the Wrangell–St. Elias, and on the west and south by the Chugach Mountains. For the last two million years, sheets of ice periodically covered most of the great Copper River Basin. It was an unsettled time, with series of warm, interglacial events sandwiched between cold, glacial periods. Great rivers of ice slowly advanced north from the Chugach Mountains to fill the Copper River Basin, then retreated, depositing huge amounts of rock rubble and silt in the basin.

There was no Copper River Delta.

About 12,000 years ago, the most recent interglacial period began. Melting ice from retreating glaciers slowly filled the basin, forming Lake Ahtna, over fifty miles wide. Brimming full, water pressed relentlessly against the ice walls of the basin. About 10,000 years ago, water pressure floated the ice that had been plugging one of the valleys on the south side of the lake. As water flowed out, it began to create a tunnel under the ice. As the tunnel grew, water poured out, faster and faster. The outflowing water scoured sediment from the lake bottom and began to carve the underlying bedrock as well as the overlying ice. After this cataclysmic lake dumping, the ice plug resettled and the lake slowly began to refill. This process repeated itself for a thousand years until the great glaciers began retreating to their present termini, about 9,000 years ago.

Living legends, passed on through generations of Native people, speak of a time when the Copper River, or *Ahtna,* as it was first known, flowed under

an ice mass that once filled the Copper River valley. As the glaciers retreated, the river eventually cut through the ice mass. The chalky brown river, loaded with silt, passed through the Chugach Mountains and immediately fanned out on the broad continental shelf. There, the river dropped its silt load, building, over time, a layer of sediment over 600 feet (185 m) deep and creating the Copper River Delta.

COPPER RIVER DRAINAGE TODAY

The Copper River's headwaters lie within the Alaska Range to the north, the Wrangell–St. Elias Mountains to the southeast, and the coastal Chugach Mountains. To the east, Mount Logan and Mount St. Elias, both in the river drainage, rise nearly 20,000 feet (6,000 m) above the nearby coast. Seven other volcanic peaks of the Wrangell–St. Elias Mountains protrude over 16,000 feet (5,000 m), making this the tallest coastal range in the world. Some of the headwaters drain a small

[28] *Delta haystacks*
'Haystacks' are the tops of old mountains buried by over 600 feet (185 m) of silt, deposited by the Copper River to form the Delta during the present interglacial period.

(PHOTO: PAT AND ROSEMARIE KEOUGH)

interior remnant of the Pleistocene ice sheet, while others drain part of the vast Bering Glacier–Bagley Icefield complex. The Bering Glacier alone, at over 2,300 square miles (6,000 km²), encompasses an area nearly double that of the State of Rhode Island and is the largest glacier in the world outside of Greenland and the polar regions. In all, thirteen major tributaries drain the 24,500 square mile (65,000 km) Copper Basin, an area roughly the size of the State of West Virginia.

Passing out of the Copper Basin, the entire river races through mile-deep canyons, then descends between the Childs and Miles Glaciers where they flow out of the Chugach Mountains. Where the Copper flows fast, boulders bounce and tumble along the riverbed, rumbling like distant thunder. The muddy, silt-laden water rasps and hisses. When the river slows, fanning out on the gently sloping Delta, sandbars pepper the waterway, increasing in size as the river spreads to over twenty-five-miles wide where it meets the sea. Drainages from other nearby glacier systems on the south side of the Chugach Mountains – the Bering, Martin River, Sherman, Sheridan and Scott – also dump their sediment loads where the mountains meet the sea. These smaller delta systems overlap the Copper River Delta, combining to create the largest contiguous wetlands on the Pacific coast of North America.

In several places on the Delta, rocky outcrops of mountain ridges protrude onto and offshore of sandy coastal beaches. On the eastern side of the Delta, near the Bering River, the Ragged Mountains meet the sea with multilayered rock, twisted and upheaved by an underlying fault system. Like giant stepping stones, bits of the mountain ridge extend offshore and are known as the Martin, Kanak, Wingham and Kayak Islands. The defiant bulge of rock on the western end of Kayak Island is Cape St. Elias, a remnant of a lava core that plugged an extinct volcano. This ridge extends underwater a distance offshore, marking the most seaward reach

of the Raggeds. Just west of the Copper River, several incongruous, forested rock islands or "haystacks" rear up on the open Delta. These haystacks are the tops of older mountains buried by a history of cataclysmic lake dumpings and continuous deposition of river sediment. On the western side of the Delta, an oblong lobe of rock extends in a southwest direction from the Chugach Mountains. The Heney Range portion of the lobe forms the western boundary of the Copper River Delta, while immediately across Orca Inlet, Hawkins and Hinchinbrook Islands form the eastern border of Prince William Sound, a deeply glaciated fjord system.

TECTONIC FORCES TODAY

The Delta is in one of the most tectonically active regions in the world. The uneasy movement of Earth's tectonic plates breeds violent, sometimes major, earthquakes that rock this region annually. The Pacific Plate continues to subduct under the continental margin at an extremely shallow angle.

[30] *Aerial view - upriver*

The ice of winter slowly gives way to spring. The Copper is a glacial river, rising each spring as the glaciers in the interior start to melt, and shrinking each fall with freeze-up.

(PHOTO: YSBRAND BROUWERS)

< [29] *Aerial view downriver*

The Copper is one of the world's most sediment-laden rivers. An aerial view downriver to the sea.

(PHOTO: YSBRAND BROUWERS)

Where these two plates collide obliquely, such as along the coast to the east and south of the Delta, they lurch and slide past each other, generating earthquakes through this strike-slip, stick-release motion. Two severe shocks (Richter scale 6.9 and 7.5) in late 1987 off the Yakatat coast (east of the Delta) were caused by such activity, as are most of the smaller annual earthquakes. Where two plates meet head-on at an ocean trench, such as off the Delta and in Prince William Sound, tremendous pressure can build when the plates get stuck, then suddenly release, as in the Good Friday earthquake of March 29, 1964, which was the largest temblor ever recorded in North America (Richter scale 9.2). This earthquake, centered in northwest Prince William Sound, distorted over 100,000 square miles (260,000 km²) of southcentral Alaska and altered thousands of miles of coastline throughout this region. Much of the Delta was raised 6 to 10 feet (2-3 m), and part of the Sound was thrust as high as 37 feet (11 m) when a giant section of the continental plate pivoted, lurching up and over the oceanic plate.

A complex series of faults underlies this region, marking areas of brittle or weak rock, which shatter under the tremendous pressures accompanying plate collision and subduction. Movement of tectonic plates can displace sections of Earth's crust up, down, or sideways along fault lines. For example, Chenega Island in Prince William Sound was thrust 55 feet to the south during the 1964 earthquake. Plate movement can also activate volcanoes in the region when light magma, created within the subduction zone, rises to the surface, or shifting plates can cut off the flow of magma and deactivate volcanoes. For example, the activity level of volcanoes in the Wrangell–St. Elias Mountains, all dormant now except for Mount Wrangell, is probably related to subduction of the Pacific Plate into the Aleutian Trench offshore of the Delta.

CLIMATOLOGY AND TOPOGRAPHY

INTERACTIONS BETWEEN THE SEA, THE MOUNTAINS, AND THE RIVER

Annually from October through March, a semi-permanent low-pressure trough, the Aleutian Low, forms over the Gulf of Alaska. This trough is a region of intense cyclogenesis, or storm formation. During these winter months, more lows are found in the Gulf of Alaska than in any other part of the northern hemisphere: the region averages one storm every five days. The low-pressure storms track northeastward across the gulf and slam into the North Gulf coastal mountains of Alaska, which open like a well-worn catcher's mitt to the sea.

Forced upward by the extreme relief of the coastal Chugach and Wrangell–St. Elias Mountains, the warm moist air chills rapidly, dumping some of the highest average snowfall in the northern hemisphere. According to data published by the National Weather Service, snowfall averages over 800 inches (2,032 cm) per year! Record snow accumulation at Thompson Pass in the Chugach Mountains near Valdez, northwest of the Delta, during the winter of 1952-53 was 974.5 inches (2,475 cm). Rainfall in the mountains is equally impressive, averaging about 240 inches (610 cm) per year, compared with an annual average of 100 to 160 inches (254 to 406 cm) along the coast, including in Cordova.

High in the mountains rimming the headwaters of the Copper, snow and rain are rapidly converted to ice by tremendous pressure from high annual precipitation. (In dry polar regions, this process takes hundreds of years.) The ice fields formed by the interaction between the storm fronts and the coastal topography feed the world's largest temperate glaciers. Temperate glaciers have an internal temperature near freezing and slide freely over their bedrock. Time-lapse photographs of temperate glaciers, sped forward, reveal them to be flowing like rivers. As this plastic ice glides downhill, it

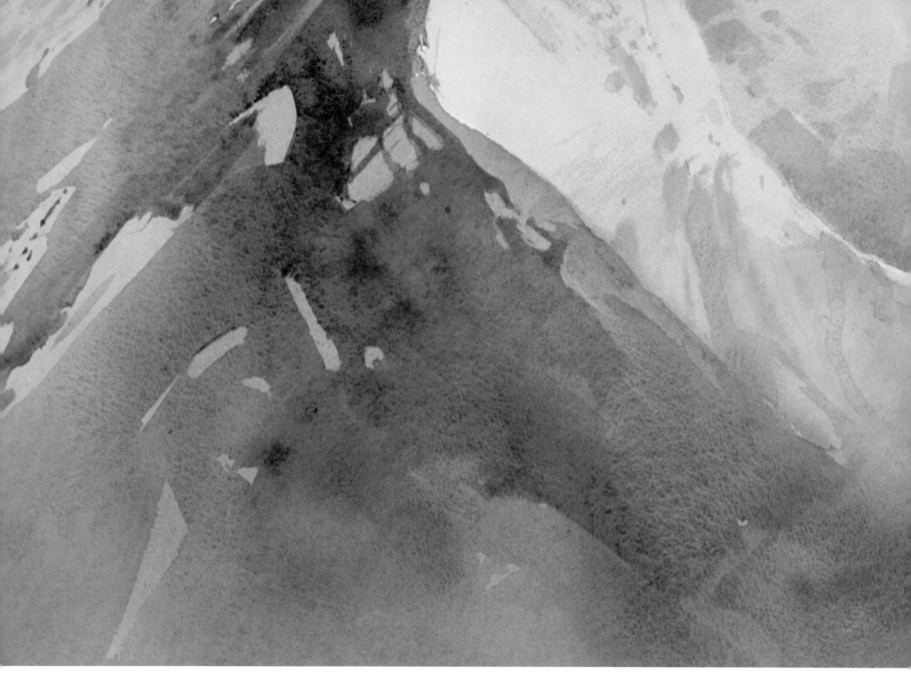

< [32] **David Barker**
Ridge
Watercolor, 37 x 55 cm
*"In the month of July the
summer night sun strikes
this north face. The south-
facing snow has already
melted to bare rock. It is
here in the mountain tops
that the Delta begins its formation."*

tears away the rock, carrying everything from fine glacial silt to huge boulders into the river.

Many of the glacier-fed tributaries draining into the Copper have glacial outburst lakes, or *jokul-haups,* the Icelandic word used by glaciologists. These lakes periodically dump their waters, like the former Lake Ahtna, but on a smaller scale. They form in valleys along the sides of glaciers where the ice acts as a temporary dam. The lake level rises until finally the tremendous pressure lifts the floor of the glacier, blowing out the ice dam and suddenly releasing all the pent-up water and sediment. Periodic outbursts (about every six or seven years for Van Cleve Lake, annually for some others) have increased the flow of the Copper by thirty to fifty percent overnight. Drainage conditions on the Delta fluctuate wildly in response to episodic flooding, seasonal melt of the snowfields and glaciers, heavy rainfall, and large and seasonally varying tides.

[33] **David Barker**
Ice Field
Watercolor, 37 x 54 cm
*"In a small airplane flying
at four thousand feet we
circled inside a great snow-
filled bowl spilling over the
heads of two glaciers."*

[34] **Todd Sherman**
Copper River from Bridge
Acrylic on paper, 54 x 74 cm
Childs Glacier, visible upper right, is the last barrier to the river as it breaks free from the mountains and enters the Delta.

[35] *Todd Sherman at work*

(PHOTO: YSBRAND BROUWERS)

[36] **Todd Sherman**
Copper River Delta
Oil on board, 17.5 x 24 cm

The annual sediment load of the Copper River is estimated at 80 million metric tons: this is over one-third the annual sediment load carried by the Mississippi River, which drains an area forty times the size of the Copper's watershed. During the peak summer floods, measuring half a million cubic feet (140,000 ds) of waterflow per second, the Copper carries about a million metric tons of sediment per day. For its size, the Copper is one of the world's most sediment-laden rivers.

A series of large barrier islands form offshore of the Delta as a strong nearshore ocean current push-es the sediment load westward. These sandy islands stretch for more than seventy miles along the coast, constantly shifting and being reshaped in response to strong storms, prevailing currents, wind, variable sediment load, and the movements of the river itself, tossing restlessly in its bed, one year draining more to the east, the next to the west.

< [37] **Bruce Pearson**
Delta Debris
Mixed media, 67 x 110 cm
Driftwood is carried down the
Copper and other rivers by
summer floods and deposited
on the Delta sandbars.

[38] **Siegfried Woldhek**
Rufous Hummingbird
Watercolor, 26 x 36 cm
The tiny rufous humming-
bird, smallest of the Delta's
migratory creatures, is a
symbol of how remarkable
and vulnerable is life on the
Delta, a thin strip of greenery
between the ocean and the ice.

Alaska, Cordova
8.76

This section of coast off the Copper River Delta is one of the few areas in the United States not covered by nautical charts: it would be useless – even dangerously misleading – to map these shifting sands at any one point in time.

The Copper River watershed forms a funnel between the dry, subarctic continental climate of the Copper Basin and the warm, moist, maritime climate of the Delta. While the North Gulf coast lies in the powerful grip of the Aleutian Low for six long months, this low-pressure area starts to loosen its hold in April, and by late May has usually dissipated, pushed from the gulf by the North Pacific High, which prevails into August. Throughout the year, steep pressure gradients between the interior and the coast are converted into wind. Pressure differences are greatest (and, therefore, winds are strongest) during the winter months, because of the Aleutian Low over the ocean and the equally strong Polar High over the continent.

The adiabatic winds of the valley are legendary. Blowing down the valley in winter (offshore) and up the valley in summer (onshore), winds of 50 to 70 miles (81 to 113 km) per hour occur every month of the year, particularly near the Bremner River mouth. Winds well in excess of 90 miles (145 km) per hour are not uncommon during the winter. The *average* daily wind speed recorded in December 1909 at the Million Dollar Bridge, where the river exits from the mountains into the Delta, was 59 miles (95 km) per hour! When the Copper River & Northwestern Railway operated, from 1911 through 1938, the railroad crews hung a heavy ball and chain at 'Flag Point' (Mile 27 bridge). If the ball and chain stuck straight out in the wind, the trains did not cross the bridge. Today, road maintenance crews working late into the fall sometimes return their heavy equipment to Cordova with the paint completely sandblasted off the windward side of the vehicles. The original 7 x 7 inch (18 x 18 cm) wood guardrail pilings on the bridge at Mile 27 are now less than 4 inches (10 cm) thick in places,

sculpted and eroded along the wood grain by the winds. Local pilots often have to ascend to 3,000 feet to avoid the engine-choking dust storms in the lower Copper River valley. Through time, these silt-laced winds have built the great sand dunes in the valley and the upper Delta.

< [39] **Vadim Gorbatov**
Raft Trip
Watercolor, 18 x 28 cm
Following the path of creation
from glacier through river to
Delta, David Grimes, Ysbrand
Brouwers and Vadim Gorbatov
descend the icy Sheridan River
down into the summer-green
Delta. Apparently the artist
remembers a particular wave
that greeted him.

[40] **Glacier Lake and Raft**
Rowing across the lake in
front of Sheridan Glacier
(PHOTO: PAT AND ROSEMARIE KEOUGH)

[43] **Juan Varela Simó**
Wading Moose
Watercolor, 29 x 39 cm

[44] **Victor Bakhtin**
Hook Point
Gouache, 29 x 62.5 cm
*Harbor seals on the rugged
outer coast of Hinchinbrook
Island, the far western boundary of the Copper River Delta.
Victor was so engrossed in
capturing this scene that he
didn't notice the brown bear
that ambled up behind him,
sat awhile to watch him paint,
then strolled off down the
beach, leaving large tracks in
the wet sand!*

[45] **Vadim Gorbatov**
Bald Eagle over the Copper River Delta

Watercolor, 54 x 57 cm
Symbol of North America, the bald eagle is no longer common throughout its former range in the United States except in Alaska, where habitat remains plentiful. These residents of the Delta concentrate by the dozens in fall near streams where salmon are spawning.

THE LIVING LAYER:
THE ECOSYSTEM OF THE COPPER RIVER DELTA

Millions of migrants fly or swim their way back to the Copper River Delta each year. The Delta is one of Earth's biological Edens, a highly productive ecosystem dependent upon a complex and intricate cycling of nutrients from alpine tundra and glaciers of the headwaters, through the forests of the upper Copper Basin and the Delta uplands, into the twisting maze of water channels and spongelike wetlands of the Delta, and beyond into the nearshore coastal marine waters – and back upstream again with the salmon and the birds.

The multiple, interdependent habitats of the Delta are separated here into four main categories for ease of discussion. *Saltwater habitats* include everything washed by the sea (or even a hint of the sea) such as coastal marine waters, barrier islands, rocky shores and sea stacks, beach and sand tidal flats, and salt and brackish estuaries. *Freshwater habitats* include low elevation streams, lakes, freshwater sloughs, bogs, wetlands, and moist meadows. *Forest habitats* include Sitka spruce–hemlock forests, cottonwood forests, shrub thickets, forested streams, and peat bogs of 'muskeg,' a wonderfully springy peat mat of Sphagnum moss carpeted with tiny plants. *Alpine tundra habitat* is the area above treeline, including tundra, high-elevation meadows and streams, and rocky slopes.

Each of the four main habitat types is distinct yet interwoven by the seasonal movements of inhabitants. There is no beginning nor end: life on the Delta follows a cycle of seasons. In early spring, life quickens first in the coastal marine waters when the hooligan fish (eulachon) return to spawn. Swans, geese, and some ducks begin to arrive about the same time as the hooligan. Within a month, there is a surge of shorebirds and salmon from the outer islands and tidal flats into the wetlands.

Steadily lengthening daylight hours rapidly green the meadows and forests and, suddenly, life is vibrant and pulsing across the Delta.

Let us merge with this cycle of life, on the coast in April when the hooligan and waterfowl return.[1]

SALT WATER HABITATS

APRIL

In April, outer coastal beaches, swept free of ice and snow by waves and tides, are gathering places for overwintering birds and mammals. On rare calm days, mixed flocks of glaucous-winged gulls, mew gulls, and black-legged kittiwakes venture to the outer beaches of barrier islands and tidal flats to search for starfish, crabs, clams, and bits of kelp and debris that wash ashore during fierce spring storms. Gregarious northwestern crows, never more than a few hundred yards inland, scavenge alongside the gulls.

Tidal sloughs and streams are locked in ice. The ice starts to buckle and break into large pans and cakes where the sea water pushes up into the river channels and sloughs on spring flood tides. Some of the gulls and kittiwakes, quick to notice these changes, gather in these areas of broken ice to wait.

Nearshore, like underwater rockets, Dall porpoise and Steller sea lions chase the first hooligan of spring. The silver-bodied, fat-rich hooligan run the gauntlet, swimming upriver in large schools. Crowds of sharp-eyed gulls wheel and cry in raucous celebration, foretelling the bounty of spring. Screaming overhead, they drop down to pluck fish from the icy waters. In turn, hungry bald eagles swoop on the mass of gulls, ending the winter famine in their own way. The dark heads of harbor seals pop up in the midst of the surface-skimming

[47] *Sockeye salmon, spawning phase*

(PHOTO: PAT AND ROSEMARIE KEOUGH)

[48] *Autumn leaves and frost upriver*

(PHOTO: DAVID GRIMES)

< [46] *Deer cabbage, Hawkins Island*

"Deer cabbage in September on Hawkins Island, across Orca Inlet from Cordova."

(PHOTO: DAVID GRIMES)

lakes and wetlands, where they grub for roots and bulbs of aquatic plants, duskys and other geese pause on the estuaries and tidal flats to feed hungrily on tender new shoots of sea arrowgrass and goosetongue.

Sea ducks congregate by the hundreds in sheltered coastal bays and lees of islands in Orca Inlet. Black, white-winged, and surf scoters dive for blue mussels, clams, snails, and small crustaceans. Brilliantly marked harlequin ducks raft near rocky islets in Orca Inlet and pluck intertidal creatures from underwater crevices. Oldsquaws, Barrow's and common goldeneyes, and buffleheads perform elaborate displays as they court their mates with the coming of spring.

From mid-April through May, migrating red-throated loons followed by common loons arrive in waves in nearshore waters, joining over-wintering yellow-billed loons. Their haunting tremolo calls float across quiet waters. These loons fish in sheltered coastal water until the ponds and lakes of the Delta are ice-free. Pacific loons quickly pass through coastal waters to more distant breeding grounds.

Other migrants arrive, some to nest and others just to stop on their way to western or interior Alaska. Double-crested and pelagic cormorants and tufted and horned puffins return to nest on rocky outcrops, sea stacks, and islets along the outer coast. Migrating gulls and kittiwakes swell flocks of over-wintering residents. Arctic terns end their long-distance journeys of some 11,000 miles from South America at the barrier islands and lower Delta wetlands. Small groups of Aleutian terns migrate directly from the high seas to the lower Delta. Parasitic and pomarine jaegers sweep along the continental shelf: some of the parasitic jaegers breed in the open marshes and meadows of the Delta. While most sandhill cranes pass over the Delta's barrier islands in spring, small flocks stop to feed on sedge flats on the barrier islands and in the Hartney Bay estuary in Orca Inlet.

[51] *Blue with green*

(PHOTO: PAT AND ROSEMARIE KEOUGH)

[50] *Marsh with blue slough*

(PHOTO: PAT AND ROSEMARIE KEOUGH)

< [49] *Early spring,
before green*

(PHOTO: YSBRAND BROUWERS)

white gulls as the seals follow the hooligan up river channels, often far beyond tidal reaches.

Migrating birds begin to arrive in waves. Millions of waterfowl migrate northward on the Pacific flyway, funneled by glaciated mountains along Alaska's narrow coastal corridor. Swans, geese, and ducks stop to rest and refuel on the Copper River Delta before continuing westward to the Yukon-Kuskokwim Delta or northward to the Copper Basin or even to the Arctic coast. The larger waterfowl – trumpeter and less common tundra swans, dusky Canada geese and, less commonly, cackling geese and Taverner's Canada geese, snow geese, and white-fronted geese – push the raw edge of spring by arriving well before most of the ponds on the Delta are ice free. While the swans fly inland past the barrier islands to flock at the Delta's larger

MAY

Another wave of migrants sweeps through the coastal marine waters in early May. Fork-tailed storm-petrels and less common Leach's storm-petrels breed on wooded islands off Montague Island in Prince William Sound. Northern fulmars pass through the offshore area en route to more distant breeding grounds in the Barren Islands, Aleutians, and the Pribilofs. Sooty shearwaters arrive from southern latitudes to spend their 'winter' season in the offshore Gulf Coast waters. Black brant sometimes concentrate in the offshore waters, but usually bypass the Delta. Humpback and gray whales also swim through the offshore waters during the early spring and summer months.

As ice-covered wetlands melt into spring, tiny shorebirds pour onto the tidal flats, wheeling in flocks of thousands. Arriving from distant estuaries in South, Central, and North America, five to seven million shorebirds stop to feed on the superabundant food reserves hidden in the vast mud flats and salt marshes of the Copper River Delta. This enormous concentration of migrating shorebirds is one of the greatest in the world. Thirty-six different species have been recorded on the Delta, the vast majority being most of the world's population of western sandpipers and Pacific flyway's population of dunlins.

During peak migration, the most popular mud flats look perforated, bearing multiple imprints of three-toed feet and probing bills. Shorebirds have evolved different feeding habits to make full use of available resources. Bill length limits the depth to which different species can probe the sand. Bill stoutness separates species by habitat such as mud, sand, pebble, or rocky beaches. Some species prefer to feed at different stages of the tide.

Western sandpipers and sanderlings race along

[54] *Tony Angell working in clay*

(PHOTO: PAT AND ROSEMARIE KEOUGH)

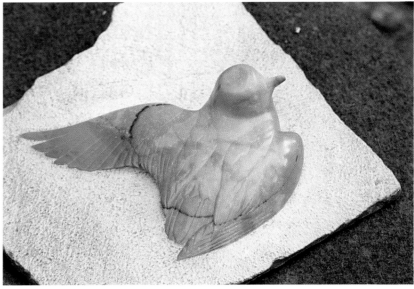

[55] **Tony Angell**
Semipalmated Plover in
Disguise

Sculpture, Sienna marble

the tideline together, but they seek out different food. Western sandpipers with their longer, more slender bills search for small pink clams, amphipods, and marine worms in the moist sand. Sanderlings swish the mud at the water's edge in search of amphipods, small shrimplike crustaceans that live near the mud/water surface and are stirred up by tidal wavelets. Like western sandpipers, dunlins prefer to feed on *Macoma* clams, but the dunlins, with their longer bills, capture slightly larger clams and hunt most actively on an ebbing tide.

On rocky beaches in Orca Inlet, stout-billed ruddy turnstones flip over pebbles, seaweed, wood, and shell debris to probe for amphipods and other invertebrates stranded at low tide. Black turnstones prefer barnacles, while cryptic-colored surfbirds search for mussels, barnacles, and snails. Rock sandpipers and wandering tattlers use their slightly longer, stout bills to probe in shallow water among pebbles and shell litter. Black oystercatchers, with their distinctive peeping cries, thrust their heavy orange bills into mussels and clams, prying the molluscs from their shells.

Some shorebirds prefer different beach zones. On the low side of the beach, shorebirds with longer legs and bills wade in shallow water to find food. So-called short-billed dowitchers (a name relative only to the long-billed dowitcher) and red knots jab their bills up and down, sometimes submerging their whole heads underwater to capture small crustaceans and molluscs. On the high side of the beach, some shorebirds feed regularly in the middle or even upper reaches of the tidal flat in the marsh grasses and sedges. Tiny least sandpipers and less common golden plovers search among the grasses for spiders, insects, worms, and crustaceans. Large, leggy whimbrels use their long, decurved bills to pull lugworms and other polychaetes from their burrows.

Dainty red-necked phalaropes have adopted a different feeding strategy altogether. Spinning like tops in shallow ponds, the birds create a vortex, like an inverted waterspout, that dredges up small aquatic insects and invertebrates. Phalaropes are the smallest true seabird and spend most of their winters in more southerly seas, migrating to the Delta and parts of Alaska to breed.

A large portion of the Pacific coast's red knots funnel through the Delta. According to the late Pete Isleib, premier Alaska birder, "when leaving the Copper River Delta, these spring migrants execute one of two spectacular migration patterns: They either depart in hundreds of small flocks (a few to about 100 birds), flying in long lines or waves only a few feet above the tide flats and Delta waters, or they form scores of large (a few hundred to well in excess of a thousand birds), compact, vocal (a mixture of raspy squeals and chattering, audible for distances in excess of one mile) flocks, flying directly north by northwest at 500 to 1000 feet (154 to 308 m) elevation" (Isleib and Kessel 1973).

The great flocks of shorebirds and waterfowl attract hungry predators. Small, swift merlins take sandpipers in fast, low horizontal flight, while northern harriers fly slower. Often hunting in pairs, parasitic jaegers prey on small sandpipers, culling one from the flock and flying it to exhaustion. Peregrine falcons stoop on ducks, shorebirds, and seabirds from high in the sky, reportedly reaching speeds in excess of 180 miles (290 km) per hour. Bald eagles also prey on migrating birds, including geese, before herring and salmon return. To confuse predators, flocks of thousands of shorebirds will flush off the beach in unison and wheel as one, alternatively flashing white underbellies and dark backs as they dodge aerial invaders.

By mid-May, when many other migrants have left the Delta, swallows start to arrive. First come the violet-green and tree swallows, usually in dribbles but sometimes in great flocks, looking like a swirling, chaotic mass of black specks released from a giant pepper shaker. Cheeping wildly, they descend en masse and flit over tidal marsh, meadow, and pond. Their enthusiasm for spring is contagious. Bringing up the tail end of migration in late May are the less common bank and barn swallows, and finally, in early June, the cliff swallows.

By the time the swallows arrive and before the flowers and grasses bloom, most birds are busy

nesting. Rocky cliffs and islands are used by a variety of sea birds and alcids. When sharing a cliff face, birds choose different areas for nest sites. The rocky Martin Islands host one of the easternmost colonies of black-legged kittiwakes in Alaska. Black-legged kittiwakes squeeze onto rocky projections too small for other birds. Gulls, puffins, and double-crested cormorants nest on grassy sites on top of these islands; gulls and cormorants above ground, puffins in burrows underneath. Double-crested cormorants and pigeon guillemots share Pinnacle Rock in Orca Inlet with black-legged kittiwakes. Pigeon guillemots nest low on the cliff in crevices under boulders and tree roots, often misted by sea spray. Pelagic cormorants nest about mid-cliff, scraping a burrow with their beak and claws. These different nesting preferences allow large numbers of birds to share very limited space.

Peregrine falcons reclaim traditional nest sites of jumbled sticks on high ledges of barren cliff outcrops. From these lonely outposts, the falcons raise their young, flying to hunt near the great gull colonies on the barrier islands or cliffs near Controller Bay. The gulls scatter like a blizzard of snow when flushed off the cliffs or beaches by hunting peregrines.

The Delta's sandy barrier islands offer a mix of beach, marsh, pond, and shrub thicket habitat. Dense colonies of glaucous-winged gulls nest inland on dry, uneven terrain behind the outermost dunes. Egg Island hosts the largest glaucous-winged gull colony in the world with some 10,000 breeding pairs. Several colonies of Aleutian terns nest across the lower Copper River Delta. Unused to disturbance, Aleutian terns do not defend their nests like the Arctic terns. They flush easily and will not renest if the first clutch of eggs is destroyed.

Nearly the entire world's population of 11,000 dusky Canada geese breeds in the Copper River Delta. The densest breeding population is in marshy areas of the barrier islands and along the

[57] **Pat McGuire**
Food Chain (triptych)
Gyotaku/collage
147 x 99 cm (x3)
Silver, or coho, salmon
chase a school of herring;
the scene created by skillful
use of gyotaku, the Japanese
art of fishprinting, and collage
techniques.
(COLLECTION OF PORT OF SEATTLE)

seaward fringe of the Delta. When the 1964 earthquake violently raised the Delta by 6 to 8 feet (2 to 2.5 m), much of the prime marsh and sedge meadow breeding area dried out. Willow, alder, and cottonwood quickly filled in the new dry land. These shrubs provide better hiding cover for terrestrial predators – brown bears, coyotes, and wolves – and better perch and nest sites for avian predators (bald eagles, ravens, gulls, and jaegers). High predation rates continue to depress the dusky population.

Marine mammals congregate to calve at this time of year. Federally listed as endangered, Steller sea lions pup in a large rookery on Kayak Island, near where naturalist Georg Steller first stepped ashore in 1741. Sea otters give birth in the shallow, protected waters of Orca Inlet and Controller Bay, on the far western and eastern ends of the Delta. Harbor seals use the quiet, inner beaches of the barrier islands and the sandy islands in the braided river Delta. Prowling pods of transient killer whales patrol the seaward channel entrances between barrier islands to seize hapless newborn seal pups swept out by strong spring ebb tides.

The harbor seals and Steller sea lions are the first to feast on the spring returns of Copper chinook (kings) and sockeye (reds) in the earliest salmon runs of any river system on the Pacific coast of North America. These salmon spend up to four years in the open ocean before navigating back thousands of miles to their streams of origin. The salmon are drawn in from the open ocean by the scent of fresh Copper River water carried seaward in distinct, muddy plumes. Bright as a silver dollar, with hues of shimmering blue or green on their backs, the sockeye surf the breakers, then follow the shore to deep channels between the barrier islands. The more powerful chinook usually swim along the ocean floor directly into channels between islands. These fish then congregate in schools in deeper portions of the channels, beginning their long journey upriver with a push from strong flood tides.

JUNE AND JULY

By mid-June, virtually all the chinook salmon are in the river system, while sockeye continue entering the freshwater systems into late July. These anadromous fish play a central role in driving the great engine of life on the Delta. By their movements, they cycle nutrients out into the open ocean and shallow nearshore sea, where the salmon are eaten by other fish, birds, and marine mammals, and into the streams and lakes of the Delta, the coastal forest, and the alpine tundra, where the fish are consumed by birds, bears, wolves, and other creatures.

With the long daylight hours, both sea and land vegetation grows rapidly. In coastal waters, dense blooms of phytoplankton, microscopic single-cell plants, drift with the currents and tides to outer, surf-swept beaches where sand dwellers such as

[58] **Susan Ogle**
Octopus
Oil on canvas, 76 x 91 cm
Renowned for its intelligence, the octopus resides in dark crevices in the rocks.

cockles, razor clams, sea pens, and some worms strain the plants from the sea. In the nearshore waters and open ocean, phytoplankton is filtered from the water and eaten by legions of zooplankton, the larval life of many marine invertebrates, and tiny animals like copepods, which spend their entire lives drifting in the currents. Copepods – billions of them – are the main link in converting phytoplankton into usable animal proteins. The tiny animals are filtered from the water by young sockeye smolts and returning adult sockeye salmon (giving sockeye their characteristic brilliant red flesh).

In the upper reaches of the estuary where tidal currents and wave action diminish, beds of eelgrass grow, their roots penetrating and stabilizing sandy or muddy channel bottoms. New eelgrass blades constantly replenish those which tear off, drift, and decay in the shallow waters, contributing to detritus, a primary source of food for filter-feeders such as cockles, little-neck, butter and other clams, mussels, sipunculid and polychaete worms, bacteria, and small amphipods, copepods, and other small crustaceans, which in turn fuel the shorebird migration. Dungeness crabs migrate inshore to molt and mate in these secluded eelgrass beds, and young crabs use the eelgrass canopy as a protected nursery where they can feed and grow. Small fish such as slender, fat-rich sand lance hide from gulls, puffins, and other sea birds in the eelgrass beds. Starry flounder, yellowfin, and rock sole, and young halibut migrate into the tidal flats and estuaries in summer to prey on young clams, sand lance, worms, and juvenile Dungeness crabs.

On the barrier islands, salt-tolerant succulent plants dominate. Many plants help stabilize sand dunes with long runners and rhizomes: beach peas, draped in reddish-blue and purple flowers, stabilize the dunes along with wild strawberries and beach cinquefoil. Beach asparagus, arrowgrass, and beach greens, all favorite foods of geese, thrive on sandy soils tucked behind the dunes, along with

sagey-smelling wormwood. Pungent lovage, with its three clusters of three leaflets, is the beach equivalent of clover. Pink flowers of searocket flag seabird nesting sites, as this plant needs high concentrations of nitrogen to thrive. Beach rye grass, commonly used by Native Americans for grass baskets, waves in shimmering fields of bluish-green leaves mixed with the large white flower clumps of wild celery and cow parsnip. Tall beach fleabane decorates sandy protected areas with its daisylike yellow flowers well into the summer. Black bears and brown bears graze upon dense clumps and mats of beach greens, at their prime in early summer, and in late July upon sweet-smelling wild strawberries.

On the rocky sea stacks, beaches, islands, and cliffs of Controller Bay, the Martin Islands, and Orca Inlet, a completely different assemblage of plants and marine creatures thrives from the surf-splashed zone to the deeper subtidal waters. Crusty black lichens coat surfaces in the splash

[60] *Colin See-Paynton by the sea*

(PHOTO: YSBRAND BROUWERS)

< [59] **Colin See-Paynton**
Kings and Loons

Wood engraving, 22 x 27 cm
Migrating red-throated loons and common loons arrive in waves from mid-April through May, joining over-wintering yellow-billed loons. These loons fish in sheltered coastal water until the ponds and lakes of the Delta are ice-free.

FLYWAY — WESTERN SANDPIPERS

[62] **Siegfried Woldhek**
Red Salmon,
Watercolor, 34 x 50 cm
Birds and salmon represent the great numbers of seasonal migrants that use the Delta. This male red salmon (Oncorhynchus nerka) is in his freshwater spawning colors, complete with arched back and enhanced jaws. (Oncorhyncus means 'hooked nose').

[61] **Tim Shields**
Shorebird Silhouette
Watercolor, 27 x 70 cm

zone. Barnacles crowd into rocky crevices in the upper intertidal zone and are hungrily sought by periwinkles, limpets, whelks, colorful sea stars, and purple sea urchins. A host of colorful sea anemones carpets the rocks from the intertidal zone to the depths.

In the turbulent intertidal zone, bladderwrack grows in profusion with pale-red sheets of dulse, sea lettuce, green rope and confetti, and the coraline red rock crust. Blue mussels form thick mats and provide cover for marine worms, snails, sea stars, chitons, amphipods, and hermit crabs.

Deeper, algae grows in patchy gardens of bull kelp, red laver, colander kelp, bushy Ahnfelt's kelp, triple rib kelp, red eyelet silk, loose color changer, and many other species. Green sea urchins graze on the kelp, while sculpins, rockfish, greenlings, and lingcod feed on smaller fish, fish eggs, and invertebrates. Octopus lurk in dark crevices, while predatory sun stars prowl over the bottom.

AUGUST AND SEPTEMBER

Although the days still have warmth, the rapidly lengthening nights chill the land and signal a change in season. By early August, the southward migration is well underway, with many adults already having passed through soon after the young fledged. The fall exodus lacks the intense frenzy of the spring return, and the migration drags on well into October. Many seabirds and shorebirds migrate in small groups, as pairs, or individually. Larger groups of waterfowl and sandhill cranes stage on the Delta when the hurricane-force winds of autumn prevent migration. Large skeins of geese, flocks of hundreds of sandhill cranes with their rusty-hinge voices, and great masses of ducks pulse southward in good weather between storms.

Meanwhile, silver (coho) salmon return to spawn in the myriad sloughs of the Delta. For the sea lions and seals, these fat salmon are the last bonanza of summer. As the silver run dwindles in late autumn,

the marine mammals move offshore into the open ocean to fish until spring.

The grasses, herbs, shrubs, and berries carpeting the barrier islands turn autumn shades of crimson, yellow, and gold. Lingering geese fatten on salt grass and beach pea pods. Voles and lemmings harvest seeds and grasses, stashing the bounty in little caches interconnected by tunnels of twisted grass stems. The marine vegetation dies back, mats of eelgrass washing out of the channels between the barrier islands, and algae tearing from rocks by the increasing storm surge. The summer fish inhabitants move offshore to deeper, calmer waters for the winter.

OCTOBER THROUGH MARCH

Winter storms roll in from the gulf with increasing regularity, bringing huge swells spawned by the mid-ocean, low-pressure Aleutian trough and by typhoons from Japan, over three thousand miles distant. The swells from these events are brought up short by the sloping, sandy shelf off the Copper River Delta, and they steepen, curl, and crash, sometimes two miles (3 km) offshore. The nearshore area becomes a frothing, chaotic jumble of sandy water. High winds whip spray and rain into stinging sheets of water, bearing inland. Even the overwintering gulls retreat to more protected waters.

The relatively protected rocky beaches and bays of Orca Inlet are used by a rich variety of life during the winter months. During severe or prolonged storms in October, fork-tailed storm-petrels will sometimes be displaced from their southward journey across the high seas into inshore waters. As winter progresses, pelagic cormorants frequent the more seaward, stormwashed beaches, while large, mixed flocks of scoters and scaups stick to the more sheltered areas of the inlet. Loons share protected bays with bufflehead, oldsquaw, mergansers, mallards, horned grebes, goldeneyes, marbled murrelets, and pigeon guillemots. Resident great

< [63] **Tim Shields**
Scatter
Watercolor, 51 x 70.5 cm
A peregrine falcon flies into a mixed flock of black-bellied plovers and sandpipers. The great flocks of shorebirds attract hungry predators such as this peregrine falcon, and also merlins, northern harriers, and parasitic jaegers.

blue herons seek the protected beaches of Observation Island, where they patiently fish in quiet pools, and the abandoned dock at Cannery Row, where they stand, statuesque, soaking up sun on rare, calm winter days. Between two and three thousand sea otters (the world's largest concentration, and about the same number as live along the entire coast of California) spend the winter in Orca Inlet, feeding on clams and a host of intertidal organisms, and hauling out on beaches to stay warm. River otters slip into the nearshore waters or forage on the beaches for snails, mussels, sea urchins, crabs, octopus, and fish.

In winters of heavy snow, many forest creatures are forced to forage on the intertidal beaches of Orca Inlet. Hungry Sitka black-tailed deer feed on kelp washed up by storms. Eagles, ravens, crows, magpies, and gulls use the beaches as primary foraging sites and quickly render any deer or sea otter carcasses. Small wintering flocks of rock sandpipers and sanderlings move about the rocky shores of the inlet.

Life in the nearshore marine waters is relatively subdued during the dark and stormy winter months – until spring, when the hooligan return.

FRESHWATER HABITATS

APRIL

In the shallowest ponds of the Delta, the ice slowly turns mushy and opaque as it loosens its grip of the bank. Then one day, the ice is gone and an inky-dark water surface reflects the sky. Overwintering trumpeter swans are among the first to spot the change. For most of the winter these swans, which choose not to migrate, are confined to the mouth of Eyak Lake, where it spills over a weir into Eyak River. The moving water usually prevents this small area from freezing, and more then one hundred swans concentrate here to feed on roots of aquatic plants. During the winter and early spring, the deserted lake echoes with loud claps from splashing webbed feet as the large-bodied trumpeter swans race across the water surface to take flight. As daylight lengthens, roving swans scout the Delta daily for dark-water ponds and settle quickly into these areas to feed and establish nesting territory. These overwintering birds – the resident breeders – are the first to initiate nests.

By early April, shallow open ponds start to fill with arriving bird migrants. Even the smallest ponds are accessible to mallards, gadwalls, green-winged teals, widgeons, northern pintails, and shovelers, all able to spring from the surface into flight when frightened by predators. Red-throated loons move into shallow coastal ponds to nest, flying to nearshore waters to feed. As the larger ponds melt and seams of open water or leads ring the shore of Eyak Lake, the more heavy-bodied ring-necked ducks, greater scaup, Barrow's and common goldeneye, common and red-breasted mergansers, bufflehead, and rarer canvasback ducks move into the Delta. These are diving ducks, with legs set farther back on their bodies, so they need a running start on water for takeoffs.

Dusky Canada geese pour into the Delta, frequenting last year's familiar sandbars, marshy pond edges, and coastal meadows as they gather in large flocks to graze on new grasses. Separate, smaller groups of cackling Canada geese, a few white-fronted geese, and, occasionally, snow geese forage intently in marsh and meadow as they refuel for the continuing migration.

Meanwhile, far upstream, cutthroat and small numbers of steelhead trout spawn in ice-free gravel riffles in the middle and upper sections of the Delta's river systems. After spawning, sea-run steelhead migrate downstream to the ocean, slowly reabsorbing their rainbow spawning hues as they pass into salt water. While steelhead range widely across the ocean, anadromous cutthroat stay in nearshore waters and estuaries. Resident trout move back into deeper lakes for the rest of the

< [64] **Vadim Gorbatov**
Rufous Hummingbird and Fireweed
Watercolor, 64 x 66 cm
Fireweed is pictured here in early summer. The buds on the stem bloom in sequence from bottom to top: "When fireweed blooms at the top, summer is over."
(Alaska saying).

[65] **Tim Shields**

Skunk Cabbage

Watercolor, 48 x 68 cm

One of the first flowers of spring, this plant grows in swampy areas of the forest. The huge green leaves with their skunky aroma come later.

[66] **Todd Sherman**

Rain Forest above Power Creek

Acrylic on paper, 57 x 55 cm

year. Elusive and even less-common rainbow trout spawn from April into July in gravel along lakeshores and in clearwater streams. A genetically plastic and readily adaptable species, rainbow trout were once, thousands of years ago, a freshwater fish, confined to landlocked lakes, but when lakes overflowed during seasons with heavy rain, these trout spilled over into river systems and gradually evolved into the seagoing steelhead trout. These fish expanded their range as the glaciers retreated, eventually establishing themselves on the Copper River Delta. Hybrid rainbow and cutthroat exist on the Delta.

MAY

Well before the brown and rusty winter tapestry changes to the soft greens of spring, ponds fairly explode with activity. Pairs of trumpeter swans display atop beaver lodges and mound nests, spreading, raising, and quivering their wings, bobbing their long-serpentine necks, and trumpeting loudly. Shy pairs of horned grebes rear up from shallow reedy water and rush across the surface on webbed toes, bearing weeds in their beaks and flaring, cobralike, their yellow head feathers in elaborate ceremonies. Ducks splash and churn the water's surface, stretching necks, flexing wings, tossing heads, shaking tails, jerking bills, and whirling and bobbing in complex displays. Arctic terns, masters of flight, sweep overhead, now low, now high, twisting and turning their white bodies in conspicuous aerial courtship flights. Parasitic jaegers, fish pirates and nest raiders, fly swiftly in synchronized flights over the Delta, their dark bodies harshly outlined against the snow fields of the Chugach range.

The excited sounds of chittering terns, gabbling, quacking ducks, chuckling, mewing gulls, trumpeting swans and honking geese, blend with the eerie winnowing of snipes diving over nearby meadows – a lusty spring cacophony fills the lengthening daylight hours with nonstop music.

By mid-May, usually even before the breakup of Copper River ice, adult chinook and sockeye salmon begin to converge in the lowest waterways of the Delta. As adult salmon make their way up the myriad sloughs of the Delta, they are plucked from the water by bald eagles and scooped up by brown bears and wolves, which switch readily from preying on feisty moose to fat-rich salmon. Harbor seals, too, pursue the salmon throughout the tidal reaches of most sloughs and streams, and up the Copper River into Miles Lake and beyond, twenty miles and more away from the sea.

Multiple generations of resident and anadromous fish are found throughout river systems and lakes in the Delta in early spring. Adult salmon moving upstream through the lower part of the Delta cross paths with juvenile salmon and juvenile Dolly Varden moving downstream and out to sea. (While salmon spend two to four years maturing on the high seas, Dolly Varden return every fall to overwinter in lakes.) Meanwhile, younger generations of salmon and Dolly Varden fry float downstream, tail first, from the upper reaches of streams where they emerged from gravel redds (nests) dug by adults the previous fall. Young sockeye and chinook make their way into quiet lakes or ponds, where they spend one to three years feeding on tiny floating zooplankton such as cladocerans (water fleas), ostracods, and copepods, and aquatic insects and amphipods living in the bottom sediments. Young Dollys spend three to six years in streams, hiding under stones or logs and feeding off insects and other creatures in the stream sediments. Most of the coho (silver salmon) fry stay within the streams, but some make their way downstream to estuarine ponds where they spend the summer before migrating back to quiet freshwater ponds in the fall. Unwary fry fall victim to mergansers, goldeneyes, Arctic terns, belted kingfishers, and loons.

There are other, less-conspicuous signs of spring. Along the edge of streams, Sitka and feltleaf

ARTIST: JUAN VARELA SIMÓ

"The light. This is a permanent remembrance of my stay in Alaska.

Or rather, the quality of the light. When you come from a southern country like Spain, summer is the time of year when painting is an early morning and evening activity. At midday the blistering sun flattens the landscape and makes nearly impossible working outside the studio. In Alaska every moment is the right time to paint. The glaciers shine like a jewel in the distance against the cloudy sky. And the constant light of summer, every hour of the day and night, softly bathing the landscape and revealing the tiniest detail of every leaf and every piece of stone.

Secondly, the silence. The dense forests are a true description of the word silence. Just like entering a European cathedral, one feels overwhelmed by the high, commanding tree trunks, like the columns of a church. The thick vegetation leaves just a glance of the next bend in the track, where you always fear to meet the mighty grizzly. The light again comes from the treetops as if sifted by the polychromous windows of the cathedral.

And in the middle of all, the sound. The splash of a bald eagle catching a salmon in Power Creek, the thunder of another piece of glacier ending its way along the ice tongue; the sudden chat of a Steller's jay. Trying to catch that sense of wilderness - this is a task for a lifetime. How could I draw or paint my impressions in just a week?"

<< [67] **John Paige**
Spruce Forest
Watercolor, 37 x 27 cm

< [68] **John Paige**
Spruce Forest II
Watercolor, 37 x 27 cm

Trumpeter Swan, adult, heavy iron staining on head, amongst sedges bordering a pool
Copper River Delta, Alaska
5th July 95

willow and the dominant Barclay willow are among the first shrubs to shrug off winter and offer smooth catkins to spring. Soon the dark, shallow waters of ponds are pierced by stalks of bright-green mare's-tail and the sweet-smelling, pinkish-white flower spikes of buckbean. In marshes and the margins of ponds, swordlike leaves of wild irises rise among bright-yellow marsh marigolds and stiff, narrow cotton-grass. By late May, frigid colts-foot blooms in loose clusters of pinkish-white flowers from stout, hairy stalks.

Meanwhile, in a world of its own making, a frigid glacial wind sweeps down the still-frozen Copper River valley, keeping spring at bay three to four weeks longer than most of the peripheral Delta. Farther upriver, the steadily increasing daylight and warmth work to the river's advantage. Flush with meltwater from its headwaters and tributaries, the Copper grinds away at its ice encasement. Usually in mid- to late May, the ice in the main channel of the Copper River gives way, rupturing the winter's stillness. Winter is flushed downstream as popping, grating, creaking, whooshing cakes and pans of ice bump and grind their way through the main river channel at Kokinhenik Bar and float out to the gulf.

JUNE AND JULY

In the nearly endless daylight of early June, the wetlands and ponds grow quieter. The great stream of migrants has moved on to distant breeding grounds across Alaska, while the birds staying on the Delta seek seclusion as they brood their clutches. Trumpeter swans remain as one of the more visible summer residents, their white plumage visible from afar and in stark contrast to vibrant shades of green throughout the Delta. The Copper River Delta hosts the largest breeding gathering of North America's largest bird: over 100 pairs of these large, graceful white swans nest in open ponds scattered across the wetlands. After the puffy gray cygnets hatch, the adults take their families into

< [69] **Keith Brockie**
Trumpeter Swan, Portrait
Watercolor, 40.5 x 29 cm
"Sketched from one of a pair of trumpeter swans on a roadside pool.... Its head is stained russet from minerals and rotting vegetation in the substrate beneath the water."

the reeds and grasses bordering the ponds, and the group becomes less conspicuous. The territorial swans tolerate little company in their ponds, but belted kingfishers and Arctic terns swoop in to dive for the ubiquitous sticklebacks and other small fish. In the airspace overhead, violet-green and tree swallows dart in highly erratic flight with civil bluet and green darner dragonflies to catch mosquitoes, black flies, and other insects.

By June, tiny black tadpoles of the uncommon western toad sometimes can be found wriggling in ponds. These amphibians waste no time once they emerge from hibernation in duff-filled depressions in the wetlands. As tadpoles rapidly develop in the short summer, adults wander throughout the grasslands, muskeg, and tundra, snapping up mosquitoes, flies, and other insects. The soft cluckings of the toad are usually mistaken for bird calls.

During these short summer months, the wetlands are a nursery for young birds. Fluffy goslings and ducklings bunch together or string out in long lines to follow adults, while precocial shorebird chicks quickly learn to survive on their own. In wetter areas of the marsh, short-billed dowitchers and least sandpipers forage for aquatic insects and worms to feed their young. In drier areas of marshes and sedge meadows, a variety of sandpipers, passerines, and some birds of prey nest on grassy tussocks or in shallow depressions of sphagnum moss, concealed with twisted grasses or overhanging vegetation. Savannah sparrows, Lincoln's sparrows, and least sandpipers distract predators away from nests by running, mouselike, along the ground. Song sparrows and fox sparrows voice nasal chirps from grassy tussocks and shrub thickets bordering marshes as they forage for their young. From flimsy ground nests of sticks and grass, northern harriers and short-eared owls patrol marsh and meadow in low, searching flight for small rodents.

The smallest mammals of the Delta – shrews, voles, and lemmings – play an enormous role in

> [70] **Keith Brockie**
> *Delta Flowers*

Watercolor, 45.5 x 30.5 cm
"Luxurious marsh vegetation behind Softuk cabin - it was nice to see lupines growing in their natural habitat."

Mile 48 from Cordova Alaska   July 95

converting the plants and insects they devour into animal proteins. They in turn provide a major food for a variety of hawks, owls, coyotes, weasels, minks, and martens. These small mammals lead short, energetic lives. By June, most have already produced one or two broods of young. Tiny dusky shrews, their rapid metabolism driving a voracious appetite, constantly hunt for insects along marsh edges. Northern water shrews, adapted for swimming in cold water, plunge into streams and bogs in pursuit of midges, dragonfly larvae, diving beetles, caddisflies, and leeches. Meadow voles use narrow runways through matted grasses along stream banks to gather grasses, bulbs, and bark for their young nestled in burrows beneath the bank. Relatively rare northern bog lemmings and brown lemmings have similar habits, but they prefer wet meadows and muskeg.

Medium-size, water-dependent mammals – muskrats, minks, and river otter – all flourish on the broad Delta. The tops of the conical dwellings of muskrats, woven with marsh vegetation, peak above the surface of marshes, ponds, lakes, and slow streams. Muskrats use underwater entrances and feed their kits roots, stems of pond-lilies, sedges, and grass, as well as an occasional clam, mussel, or fish. Long, slender-bodied minks slip out of their dens along streams and lake banks during quiet night hours to hunt rodents, eggs, fish, and birds. Larger river otters swim with webbed feet in agile pursuit of Dolly Varden and salmon to feed hungry pups.

Beavers are the true engineers of the Delta. Before the wetlands become a sea of sedges, grasses, and rushes, and before leaves cloak the streamside alders and willows, their handiwork is most evident. Constant tinkering by a work force of over 3,000 beavers – one of the largest concentrations in the world – reshapes drainages after a harsh hand dealt by geologic forces. Beavers need two to three feet of water to provide safety from wolves and bears. In quiet sloughs and connected ponds, bea-

[72] **Tony Angell**
Hawk Owl
Marble sculpture, life size

vers build dams of cottonwood, alder, and willow twigs carefully secured with mud, sticks, rocks, and plants. Domed lodges, used to rear young and cache food, are then constructed upstream of the dams. In streams too swift or large for dams, beavers build bank dens by burrowing into the stream bank and then heaping trees, rocks, and mud on top. As beavers clear brush and small trees, they create new food patches and nesting ground for waterfowl. Ponds created or deepened by dams serve as a nursery for young Dolly Varden, sticklebacks, and salmon. In late spring, beavers are busy with new young and, unless a critical dam is weakened by spring floods, construction activities are at an ebb.

Brackish and freshwater marshes dominate the Copper River Delta between the estuarine and upland areas. These marshes are a no-man's-land of

< [71] **Andrew Haslen**
Hawk Owl, Mile 48 from Cordova
Mixed media, 73 x 108 cm
Northern hawk owls hunt during daylight as well as at night. Most often seen perched high on a tree limb, they can usually be closely approached.

sea arrow-grass, spike bentgrass, sweetgrass, meadow barley, bluejoint, densely tufted hairgrass, gently waving sedges, and red fescue. The sweet-smelling marshes are a beautiful wash of silvery green with a purplish-reddish cast in the terminal tufts and spikelets. Well beyond the reach of the sea, these freshwater marshes, from the wet sedge meadows adjacent to the slough banks to just below the grass banks, still swell twice daily with fresh water backed up by high tides. During these times, the swollen marshes are a favorite feeding area for ducks.

Moose frequent willow-alder thickets, marshes, and shallow ponds in secluded areas of the wet-lands. Cows, their leggy calves in tow, and solitary bulls spend the summer browsing willow twigs and sedges, horsetail, pondweeds, and grasses from the bottoms of ponds. When feeding in a pond, a moose plunges its entire head underwater, emerg-ing later with a dangling mouthful of aquatic plants and a head streaming with water. They munch, perusing the pond's margins for predators before plunging back down for another mouthful. When cows are feeding in deeper water, young calves often nestle for a nap in a protected thicket along the shore.

From June well into mid-August, a continuous succession of plants bloom on the Delta. On the well-drained cutbanks above the maze of river channels and sloughs, narrow highways of flowers parallel the braided waterways for miles, the pre-dominant colors changing throughout the summer's succession of blossoms. In the marshes, stout wild sweet peas boast bright pink, sweet-smelling blossoms. Shore buttercups, bright-yellow beach cinquefoil, magenta shootingstars, dark chocolate lilies, slender and fragrant white bog-orchids, and purple tall-Jacob's-ladder grow hidden among high grasses and sedges. Poisonous water hemlock, with its bouquet-like umbels of small, white flowers, grows both in marshes and shallow lakes. On the upland side, the low-elevation marsh-

DIPPER and SALMON / POWER CREEK, CORONA, ALASKA. ANF. JULY '95. David Bennett

>> [77] **David Bennett**
Red Salmon and Wolf
Watercolor, 55 x 38 cm
Copper River Delta wolves
consume more fish than do
any others in Alaska and
perhaps the world.

> [76] **Keith Brockie**
Wolf Tracks
Watercolor, 28 x 28 cm
"Sand betrays the presence
of many unseen creatures,
and I was delighted to find
wolf tracks on the beach still
wet from the receding tide.
The knowledge that these
creatures are around gives
the spirit a real uplift."

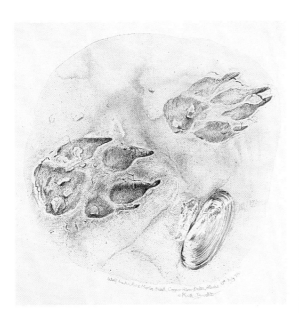

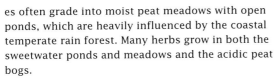

[75] **David Bennett**
Dippers, Power Creek
Watercolor, 54 x 73 cm
American dippers venture
into swift-flowing streams
where they stride along the
bottom or swim through the
current to capture aquatic
insect larvae.

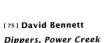

es often grade into moist peat meadows with open ponds, which are heavily influenced by the coastal temperate rain forest. Many herbs grow in both the sweetwater ponds and meadows and the acidic peat bogs.

Streamsides and gravel river bars in the Delta host somewhat different plant associations from those found in the acidic muskeg bogs. By mid-June, gravel bars are brightened with blooming magenta dwarf fireweed, wild sweet pea, and endless mats of yellow dryas. In July, the delicate spring-beauty opens its showy, pale-pink, darkly veined flowers, and yarrow blossoms into bunched white bouquets. Streamsides are an early spring tangle of purple wild iris, dainty violets, spindly pink-flowered fireweed, and trailing magenta nagoonberry blossoms. Later, brook saxifrage, goatsbeard, and daisylike, lavender Siberian aster come into bloom, to be followed, finally, by toothy triangular-leafed fleabane and sprawling wild snap-

dragon. Distinctive evergreen deer ferns, bushy, spreading lady ferns, palmate maidenhair ferns, and deciduous, narrow beech ferns grow in profusion along the banks.

By early June, tens of thousands of chinook and sockeye salmon are headed up the Copper River. Once the fish cross the seaward channel entrances, it may take them one to six more weeks to reach their natal streams, depending on the distance upstream. The river narrows where it flows in front of the Childs Glacier, and sections of the ice face frequently dislodge or calve, crashing thunderously into the river and displacing the water high up onto the opposite bank. Occasionally, salmon wash up into the forest, stranded there by these cascading waves.

At the seaward entrance, the chinook run dwindles by mid-June, as does the run of sockeye, which goes up the Copper River by late June, but 'delta spawners' – sockeye salmon that home to other streams and sloughs on the Delta – continue to enter the system well into July. As the great rivers of red fish swim upstream, they pass through beaver ponds and lakes, and rest in deep eddies and pools caught downstream of fallen logs. These areas are all prime rearing habitat for young salmon and trout. Many sockeye spawn in sandy and gravelly lake shores, where percolating groundwater oxygenates the eggs and young alevins. Some sockeye spawn in streams that are tributary to a lake where the young fish rear. These juvenile sockeye form large schools in lakes and are a major food source for fish-eating birds. Some adult sockeye spawn in streams without lakes. Usually these streams are either lake-like in nature or are adjacent to an estuary. In these systems, juveniles rear in pools and eddies in the stream or in the nearby estuary.

Sockeye and other fish are the main food source for about twenty to thirty wolves, which roam the Delta in five packs of varying sizes: in fact, Copper River Delta wolves consume more fish than do any

other packs in Alaska and perhaps the world. Wolves fully utilize the Delta, from the mud flats to the alpine tundra, and supplement their fish diet with beaver, geese, moose, an occasional goat, and some small game such as hares. Although the wolf packs on the Delta are very dynamic, there is little movement to or from areas outside of the Delta.

AUGUST AND SEPTEMBER

In early August, coho salmon and Dolly Varden return to the Delta's river systems to spawn. Spawning activities stretch into late December. For the salmon, this is a one way trip: they die soon after spawning. Dolly Varden, however, move throughout the Delta after spawning, feeding on salmon eggs, aquatic insects, beetles, leeches, and small fish of any species, including their own. Dollys, cutthroat, and rainbow trout move into lakes to overwinter during this time, before the onset of the fall floods.

As the coho and Dollys return to the Delta, the birds are heading south. (Many young birds, especially the smaller songbirds and shorebirds, fledge by late June or July and head south soon after.)

[79] **Andrea Rich**
Salmon, Power Creek
Woodcut, 20.5 x 36.5 cm
"It's an amazing thing to see. These giant fish have swum all the way from the ocean, changing color, losing their scales, all so that they can breed and die here in the shallow water in this stream by the side of the road."

< [78] **Victor Bakhtin**
Red Salmon and Bear Claws
Gouache, 12 x 19 cm

< [80] **Vadim Gorbatov**
Hummingbird, Columbine
and Red Salmon
Watercolor, 46 x 68 cm

[81] **Vadim Gorbatov**
Black-tailed Deer near
Hook Point
Watercolor, 30 x 42 cm

[83] **Tony Angell**
Spruce Grouse
Limestone sculpture, life size

< [82] **Vadim Gorbatov**
Spruce Grouse Family
Watercolor, 30 x 42.5 cm
*The forest high on the
slopes of Mt. Eyak above
Cordova teems with life.*

Arctic terns, birds of eternal summer, leave by early August on their annual migration of 22,000 miles between the Arctic and the Antarctic. Families of dusky Canada geese unite to build the great V-flocks soon winging south. Swans, ducks, and geese arrive on the Delta from distant breeding grounds in interior and western Alaska. These birds refuel on the Delta before continuing south. On clear nights between storms, the rusty-hinge voices of passing sandhill cranes float down from dark, star-studded skies ablaze with rippling green northern lights.

The marshes turn into a sea of tawny yellow, gold, and straw-white as the grasses and sedges mature. Along streams, bright-red fruits of mountain ash, elderberry, and nagoonberries sparkle from among the yellowing herbs and grasses.

OCTOBER THROUGH MARCH

The crisp, clear nights of October bend the marsh grasses with silvery hoar frost. Surfaces of shallow ponds become works of art as irregular shards of skim ice bump together and freeze, creating a crystal patchwork. The sun, at an ever-lower angle, catches heavy morning mists rising off the marshes and ponds, and the soft, yellow rays give life and depth to the swirling mists.

Ponds and lakes slowly close as the skim ice thickens, first into a clear plastic wrap that flexes and bends with pressure, then into the rigid armor

[84] **Todd Sherman**
*Vole Trails in Alpine
Muskeg*
Acrylic on paper, 28 x 45 cm
*Exposed by spring snow
melt, the muskeg is cris-
crossed by the tiny trampled
trails of voles, shrews, and
lemmings, which use these
paths under the snow to
access their winter food
caches.*

that creaks and pops with underlying water movement. Sometimes there are several false starts with heavy rainstorms melting the ice, but usually by November even Lake Eyak is locked up tight under a thick layer of ice.

Much activity goes on under the ice. Muskrats remain below the ice for long periods, feeding on cached food and submerged vegetation. They maintain a network of 'pushups,' piles of marsh vegetation covering open holes in the ice. By stopping to breathe and feed at these pushups, muskrats can take extensive outings under the marsh's icy surface. Beavers do not maintain a system of breathing holes, but they do leave their dens to retrieve food stored during the summer in underwater caches. River otters slip under the ice where gas bubbles or cracks from water movement prevent the ice from freezing. The otters dig elaborate tunnels to these fishing holes in the snow overlaying the ice.

The snowy marshes of winter reveal a road map of animal activities. Frozen sloughs and streams become travel corridors for moose, which establish a hedge line along these frozen waterways as they browse terminal shoots of alder and willow. The Delta is crisscrossed with the packed trails of these heavy-bodied animals. Twinned tracks of the weasel family indicate clearly where mink have hopped across the snow to pushups, where they slip into the water and pursue one of their favorite foods, the muskrat. Tiny tracks of ermine (the white, winter phase of shorttail weasels) reveal their high-strung nature as the prints bound along, looping back and disappearing under snow. The hops and slides of river otters look like a giant Morse code message in the snow – dot, dot, dash, dot. Oval padded tracks of snowshoe hare never lead far from a shrub thicket. Even birds leave tracks. Ravens and eagles make wing marks where they land. Willow ptarmigan, invisible in their white winter plumage except for their red combs, leave feathered tracks where they feed on willow twigs. Great horned owls create plunge pits where they

strike the snow, talons first, to seize rodents or small weasels moving about under the snow.

Meanwhile, the Copper River lies shrunken and still, a barren, frozen landscape sculpted by howling winds. Strong, bitterly cold wind from the interior of Alaska funnels down the river corridor, pushed by high pressure in interior Alaska and pulled by intense low pressure in the Gulf of Alaska. These weather conditions rule supreme until spring.

FOREST HABITATS

Home to the western hemlock and Sitka spruce, the rain forest of the upper Copper River Delta forms a unique part of an uncommon forest

land subsides, the forest retreats, leaving drowned tree skeletons to mark the limits of its advance. In poorly drained meadows, peat bogs form and limit forest growth. This occurs as heavy rainfall leaches the soil, and forest litter decays slowly in the cold climate: over time an infertile soil layer builds up which, once waterlogged, forms an impermeable layer, ultimately converting to acidic peat bogs. Along the forest margins and in areas opened by blow-down or logging, Sitka alder, a pioneer species, improves barren soil with nitrogen and organic leaf litter. Harsh subarctic conditions limit timberline to about 1,500 feet (460 m) above sea level. The only place where the rain forest can maintain its balance with wetland, muskeg, and moraine is along steep slopes – the mountains and foothills of the Chugach, Ragged, and Heney Mountains, and in the rocky haystacks protruding through the wetlands.

In different soil types along glaciated river beds and glacial moraines, the coniferous forest gives way in places to black cottonwood. Along the margins of the rain forest, between the forest and the wetland, lake, stream, muskeg, or alpine tundra, dense thickets of alder grow at higher elevations. At lower elevations, especially along sloughs, streams, and larger flood plains, shrubby Sitka willow and the larger feltleaf willow grow mixed with the alder.

APRIL

In early April the great rain forests of the Delta lie mostly quiet. Small flocks of resident birds – pine siskins, juncos, redpolls, chickadees, Steller's jays, magpies, pine grosbeaks, and white-winged crossbills – flit about the forest in pursuit of conifer seeds. Red squirrels search out their last caches of green spruce cones, cut and stored the previous summer. The snow cover turns wet and heavy as it melts, making movement difficult. After a long winter, many forest creatures have exhausted their food reserves. It is a time of waiting.

< [85] **Bruce Pearson**
Dwarf Hemlock and Rufous Hummingbird
Mixed media, 46 x 68 cm
Muskeg bogs, with gurgling pools of water and stunted mountain hemlock, are displayed like bonsai gardens. These poorly drained, acidic peat bogs limit forest growth.

system. It is part of the northernmost range of the coastal temperate rain forest that extends 2,000 miles (3,000 km) along the Pacific coast from northern California to the Kodiak Island archipelago in Alaska. Worldwide, only two to three percent of temperate forests are coastal rain forests. Half of the world's coastal temperate rain forests are on this Pacific coast. The essence of these cool, damp forests depends on copious rains produced when warm, moist oceanic air masses are forced to rise over steep coastal mountains. This forest ecosystem is largely associated with watersheds which drain directly into the sea.[3]

The subarctic rain forests of the Copper River Delta are extremely sensitive to change. When an earthquake lifts the Delta, the forest gradually expands out to drier ground in the wetland; as the

Then one day in mid-April, the songs of ruby-crowned kinglets and the sweet notes of varied thrushes ring out from deep woods as the first wave of migrants arrives. Each day the chorus swells with the cheerio of robins and the trills, warbles, whistles, and calls of warblers, finches, and juncos. The loud wing-clap of spruce grouse adds to the chorus. The forest smells of rich, moist earth. Small, tightly coiled tips of ferns start to poke through ground still saturated with snow melt. From water-filled pits and swampy areas in the uneven forest floor, the bright-yellow spathes of skunk cabbage rise furled around prominent lime-green flower spikes, while the huge green leaves with their skunky aroma come later.

In mid- to late April, brown bears emerge from dens dug into alpine slopes, first the males and later the females with cubs. The hungry bears make their way to the wetlands and Delta to forage for roots and new grasses, sedges, and horsetails. Emerging black bears stick close to the forest, feeding on freshly sprouted greens and winterkill.

MAY

In early May, shrub thickets along the margins of the spruce-hemlock rain forest tremble with movement of hermit thrushes, Wilson's warblers, and golden-crowned, Lincoln, and song sparrows. Fox sparrows scratch busily in dry leaf litter on the forest floor. In a few short weeks, these small songbirds will be nearly impossible to spot as they forage, hidden by dense new growth.

In peat bogs sprinkled throughout the spruce-hemlock forest, there are other signs of spring. Typical of poorly drained areas near spruce-hemlock forests at this latitude, the peat bogs consist of meadows, ponds, fens, and muskeg. Soon after snow melt, the muskeg is splashed with color from small, early-blooming evergreen and deciduous plants: maroon crowberry, pale-pink bog cranberry, red bearberry, pinkish-white lingonberry, and bright-pink bog blueberry. Small sundews sport

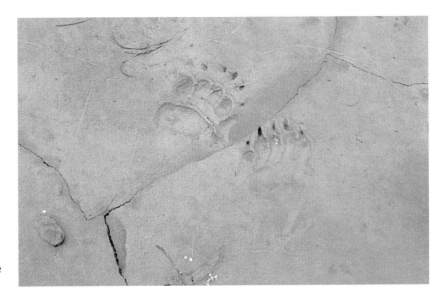

tiny white flowers and ensnare mosquitoes, midges, and gnats with their sticky leaves.

By mid- to late May, most of the migrant birds have returned and, along with the overwintering residents, are courting, mating, and incubating. Nesting activities occur in all levels of the rain forest. Usually high in the tree or on a cliff face, ravens build massive stick structures, while great horned owls and less-common merlins use abandoned nests of northwestern crows or red squirrels. Northern goshawks and sharp-shinned hawks claim major crotches near the midpoint of conifers for their large stick nests. Skillful flyers, the sharp-shinned hawks flash through relatively thick forests, snatching small birds from their perches. Pine siskin, Townsend's warblers, and occasional white-winged and red crossbills nest at different heights on horizontal branches far from the trunk. Tiny rufous hummingbirds, remarkable migrants from as far away as California, swoop aggressively at intruders to defend teensy nests tightly bound

[87] *Bear paw prints*

(PHOTO: PAT AND ROSEMARIE KEOUGH)

< [86] **Bruce Pearson**
Black Bear, Hartney Bay
Mixed media, 30 x 41.5 cm

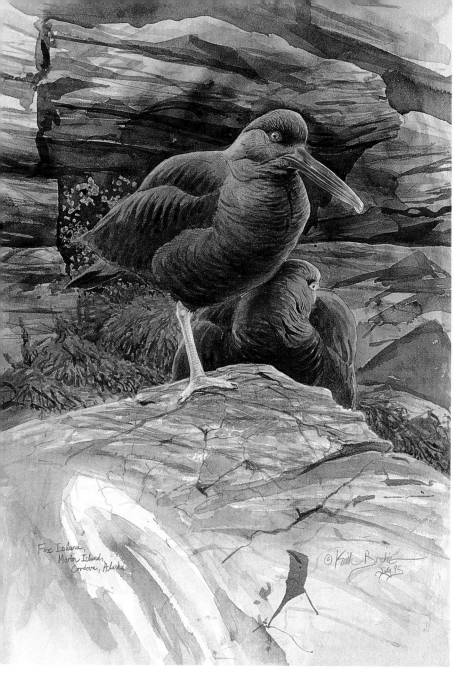

Fox Island,
Marten Islands,
Cordova, Alaska

© Keith Brockie
'95

to drooping limbs with spider's silk. Ruby and gold-en-crowned kinglets build pendant, saclike nests hung from limbs near the trunk and woven with spider's silk, lichen, and moss, while shy marbled murrelets seek shallow depressions on moss-covered branches of the largest trees. Steller's jays nest near the trunk or in the crotch of a branch. Diminutive brown creepers spiral around tree trunks during courtship and nest under loose, scaly bark on the trunk. Cavities in trees are used by several duck species: common mergansers and Barrow's goldeneyes prefer live trees, while buffle-heads and common goldeneyes use standing dead trees or 'snags.' Hermit thrushes and orange-crowned warblers craft nests hidden under litter on the forest floor, while spruce grouse lie hidden under branches or the trunk of a fallen tree.

Mixed stands of black cottonwood, mountain hemlock, and Sitka spruce, like those in the river canyon below Saddlebag Glacier or on the alluvial floodplain of the Copper River, house many of the thrushes, warblers, and sparrows found also in the rain forests, but some species prefer these more open mixed forests. Flashy black-billed magpies nest in open crowns of cottonwoods, while northern hawk owls seek out dead standing snags.

In tall spruce and hemlock along the east side of Orca Inlet, great blue herons maintain rookeries of large, flat nests of interwoven sticks. Small colonies of these birds remain year round, fishing in nearby marshes and coastal waters.

Many birds use the streams as flyways to navigate through the forest, and others rear their young in the streams or nest in the stream banks or in nearby man-made structures. Common mergansers tuck their brood under dense, overhanging stream-side vegetation when threatened and, with much splashing and ado, distract the predator away from hiding young. Bank swallows burrow into stream banks, while cliff and barn swallows prefer to plaster their muddab nests on the underside of bridges or buildings near water or open wetlands. American

< [88] **Keith Brockie**
Black Oystercatchers,
Fox Island
Watercolor, 42 x 29 cm
A pair roosting among the
dark rocks and seaweed.

dippers prefer stretches of fast water and build little huts of woven grasses on cliffs among maidenhair ferns and mosses or behind waterfalls. These small slate-colored songbirds venture into the streams, where they can stride along the bottom or swim through the current to capture aquatic insect larvae.

The coastal temperate rain forest shares a special relationship with its streams. In late May, salmon fry wriggle up through gravel redds. The gravel beds are often downstream of fallen logs, which obstruct and slow the stream flow. Conifers decay very slowly, and when trees fall into stream channels, they may remain intact for over a century. These fallen logs slow the torrential spring and fall floodwater as it rushes down steep mountain slopes, thus helping to prevent small salmon fry from washing out to sea. Needles, leaves, lichens, and wood debris rain down from the forest canopy and nourish larval forms of aquatic insects – stoneflies, caddisflies, mayflies, midges, and dragonfly nymphs. Logs and branches create quiet pools and eddies where young coho salmon spend the summer, hiding under woody debris and darting after drifting insects. Other fish such as young Dolly Varden also rear in these quiet, shaded stream pools.

By late May, pregnant cow moose seek open thickets of tall alder and willow for calving.[4] Brown bears and wolves learn where the moose concentrate, and they search these thickets for a baby moose snack. To increase the survival odds for individual young, cow moose give birth during a very short birthing season. The bears and wolves cannot capture all the vulnerable newborns at once and, within a few days, the young moose can outrun these predators.

JUNE AND JULY

In mid-June an afternoon hike along the Eyak River Trail[5] brings one the best of two worlds. Entering the forest through a tangle of salmonberry, mountain ash, elderberry, and fireweed, the trail follows the Eyak River for the first half mile, winding through a second-growth spruce-hemlock rain forest. The ground is covered with mosses, clubmosses, unfurling ferns, and blooming bunchberry, spindly pink pyrola, one-sided wintergreen, wax-flower, twistedstalk, false Solomon's-seal, Labrador tea, trailing raspberry, wild geranium, and violet, among other herbs. Waist-high understory shrubs are hard to miss in colorful spring bloom: high bush blueberry, cranberry, black currant, and

[89] **Susan Ogle**
On the Lookout
Oil on canvas, 91 x 91 cm
Dusky Canada geese protecting their young during a typical summer storm on the Delta.

[91] **John Paige**
Surf Scoters and Sea Otter
Watercolor, 25 x 34 cm

< [90] **Juan Varela Simó**
Harlequin Ducks
Watercolor, 30 x 40 cm

devil's club. Old man's beard lichen hangs in wisps from limbs and bark of conifers, while rag lichens grow in leathery flakes from fallen logs.[6] Steller's jays and red squirrels scold from the safety of tall trees.

After a half mile, the trail climbs abruptly to a sunny, rolling stretch of muskeg bog and meadow along the edge of the rain forest. The trail, two-by-eight planks laid end to end, leads on across the fragile muskeg for two miles. In drier spots of the bog, clumps of lichen-draped mountain hemlock and Sitka spruce, their growth stunted and twisted by the heavy rainfall and poor drainage, are displayed like bonsai gardens of the gods. Bathtub-size ponds of dark-brown water lie in flat depressions, while on slight slopes, smaller ponds are interconnected in stairstep fashion by little gurgling water falls. The muskeg is crisscrossed with tiny, trampled trails of voles, shrews, and lemmings, which used these paths to access winter food caches. Exposed by snow melt, the trails will soon be hidden again by new summer growth.

Rookery   1/24 '95   Andrea Rich

Showy wild irises with broad purple flowers, distinctive chocolate lilies, and deep-red western columbines flag the edges of depressed ponds with patches of lady ferns, deer ferns, and narrow beech ferns. Slightly drier areas of the bog are marked by fragrant sweet gale and Labrador tea, with its clumps of small, frilly white flowers. Varied thrushes trill from the edge of the forest. From high in conifers, stilt-legged yellowlegs watch as warblers and sparrows flit across the open meadow. High overhead, bald eagles circle in the thermals. At the edges of the meadow, one might glimpse a coyote before it fades into the forest.

In this same rain forest during the brief, twilit summer night, coyotes call from open meadows above the river bed. Timid snowshoe hares emerge from protected thickets to feed in wooded swamps and brushy areas. Great horned owls glide through the woods on silent wings, a whispered presence. Little brown myotis bats leave daytime roosts in tree cavities and caves to hunt for insects. Porcupines climb slowly and awkwardly, munching on the inner bark of spruce and hemlock. Weasels, fierce and voracious predators, search for rodents, shrews, and young hares in the woods and shrub thickets. Pine marten follow the fringes of spruce thickets along bog meadows to hunt voles.

In the early morning hours, black bears prowl along stream banks at the forest edge.

By mid-July the forest and muskeg have donned summer attire. In open woods, giant cow parsnip, wild rhubarb, and goatsbeard wave white flower heads at waist level, while bluebells, Alaska spiraea, and false hellebore bloom among others on the forest floor. Edging the open muskeg meadows, larkspur and monkshood wave purple flowers atop tall stems. Around ponds, reddish-purple marsh cinquefoil and magenta shootingstars bloom among northern green bog orchids, yellow bog star, bog saxifrage, and the fragrant flowering spikes of white bog orchid. The surfaces of ponds are nearly covered with the large floating leaves and tennis ball-size flowers of yellow pond-lilies.

Brown bears drift back from the Delta up sloughs and streams into forests where they gorge themselves on ripe salmon returning to spawn. Essentially the same species as the interior grizzly, the coastal brown bears nevertheless often attain a much larger size because of their nearly unlimited access to fat-rich salmon. Sows huff and chuff as they teach their young cubs the ancient fishing skills. When canoeing quietly down one of the many streams on the Delta, one can often round a bend and find a freshly caught salmon still flopping on the bank.

AUGUST AND SEPTEMBER

There's a saying in Alaska: When fireweed blooms at the top, summer is over. By early August,

[93] *Aerial view of kittiwake colony, Controller Bay*

(PHOTO: YSBRAND BROUWERS)

< [92] **Andrea Rich**
Rookery

Woodcut, 33 x 42.5 cm

"I spent a few nights on this small rocky island where I had come hoping to see puffins. The puffins were there but in burrows on the inaccessible slopes. Instead I saw otters and a variety of shorebirds and, when the fishing boat came to collect us, this densely populated kittiwake rookery that had not been visible from the other side."

[94] **Dylan Lewis**
Gulls
Watercolor, 20 x 30 cm

the uppermost buds on the stem are blossoming, while the lower blossoms have already changed into long, slender seed pods. In the next few weeks, the leaves and stem will turn a deep scarlet 'fire' red, topped by a mass of maroon seed pods with puffs of fluffy white seeds.

Many of the rain forest's migrant songbirds – the warblers, thrushes, sparrows, and finches – head south well before the fireweed is finished blooming. Rufous hummingbirds start leaving in late July. The emptiness is filled by gulls, ravens, crows, and eagles, which fly into the forest and congregate along salmon-spawning streams, feeding heavily on dead and dying salmon.

From a distance, the conifer forests of the Delta appear the same silvery-blue deep green throughout the year. But within the forest, the understory

and the forest floor blaze with color. The deep-green trees accentuate the red berries hanging in clumps or singly from mountain ash, elderberry, high bush cranberry, false Solomon's-seal, twisted-stalk, and salmonberry. Shrub leaves turn shades of mustard to golden yellow. Devil's club, with its spiny branches, bright-yellow leaves, and spikes of red berries, stands out among the shrubs – even the berries of the stickery plants have spines to prevent black bears from munching in complete delight. In certain places on the Delta, wild crab apple trees produce small, oblong red and yellow fruits. Pockets of muskeg meadow within the forest turn a watercolor wash of reds, greens, and yellows. Most of the evergreen and deciduous shrubs in the sedge-moss peat mat sport firm black, blue, red, or maroon berries. Stands of cotton-grass rip-

[95] **Dylan Lewis**
Resting Gull
Watercolor, 22.5 x 19 cm

ple in waves when the white tufted seed heads are stirred by wind.

In late summer and fall, mushrooms emerge from the forest floor or from rotting trunks and stumps. Mushrooms are the fruiting bodies of fungus organisms that grow unseen underground or in decaying wood. The red and maroon caps of russulas brighten the forest floor with orange and green milk mushrooms. Coral mushrooms pop out of the duff beneath spruce and hemlock along with troops of their cousins, the fairy clubs. Pale golden trumpets of funnel chanterelles and 'black' chanterelles grow in large numbers in the forest moss along with pale-orange hedgehog mushrooms (with tiny teeth in the place of gills). Shelf fungi, such as the artist's conk and brilliant yellow-orange chicken-of-the-woods, sprout from tree trunks. Little white angel wings, in overlapping clusters on the sides of fallen logs, shine in the shady forest gloom.

Some fungi share an important symbiotic relationship with hemlock and spruce. Fungi lack chlorophyll and so cannot produce their own sugars for food. In some species, the intricate weblike roots of the fungus, called *mycorrhizae,* intertwine with tree roots and grow threadlike hyphae to penetrate tiny rootlets. Trees exchange photosynthesized sugars for minerals, water, and other nutrients which the fungus absorbs from the soil. (A nitrogen-fixing bacteria lives within the fungus and provides usable nitrogen for both the fungus and tree in exchange for food.) These beneficial relationships help the rain forests grow in the poor soil conditions of the subarctic.

In mid-August, the first fall gales and storms sweep down on the Delta with high winds and driving rain. The storms of fall are not like those of winter, when rain or snow lashes with a violent vengeance from brooding, dark, leaden skies. The fall storms are tumultuous and indecisive. Billowing clouds shift in rapidly moving, grayish-white shapes and tear open in great raggedy holes to

[96] **Pat McGuire**
Rock Bottom (detail)
Gyotaku/collage, 91 x 183 cm
A fishprint: up close and personal with lingcod.

show teasing patches of bright-blue sky. Strong winds blow in gusts and fits, swirling from all compass points. On these blustery fall days, ravens and crows take time out from scavenging to play. They congregate near the tops of conifers growing near an open slope or meadow. A dozen or so of the birds launch and perform great aerial somersaults, loop-the-loops, and barrel rolls as their audience caws and croaks from bobbing tree limbs. If the day is particularly gusty and the rains hold off, the show may go on for hours.

In late September the deep bellows of bull moose in rut ring out from alder-willow thickets. Often these bellows attract nearby bulls as well as the desired cows. Male moose in prime condition weigh from 1,200 to 1,600 pounds (542–725 kg). With much thrashing of the thicket, the bulls joust, clashing their antlers together and pushing. While serious battles are rare, only the dominant bull usually mates with receptive cows.

OCTOBER THROUGH MARCH

During the fall and winter, high winds along the coast frequently blow down spruce and hemlock, which have a very shallow root system (usually no more than a foot, 0.3 m, deep). These blow-downs (not fire, as in some drier forest ecosystems) are critical for maintaining the forest's health, because the action of ripping the root wad from the organic mat aerates, mixes, and drains the surface soil layer. Haircap mosses colonize the fallen trees and their upturned tree roots, slowly breaking down the wood with their fine, hairlike *rhizoids* or rootlike feelers. Blow-downs create openings in the dense forest canopy. Flooded by light, hemlock and spruce seedlings will sprout rapidly in these areas in the spring, perpetuating the growth of the rain forest.

Like turning off a tap, streams slow and water levels drop as freeze-up claims the water as ice for the winter. Yet well into December, late runs of silver salmon still return to spawn. The gaunt, dark-red salmon circle sluggishly in spawning grounds, their bodies tattered from swimming up streams that are more rock than water and from breaking through razor-sharp skim ice. Some of the resident gulls and eagles gather near the spawning grounds to feed on salmon carcasses. Without much current to carry off waste, the spawning areas in late fall smell strongly of rotting flesh and bird excrement, but later, in the spring, nutrients leached from this organic mass will nourish insect and aquatic life and the water itself, fueling the growth of the next generation of salmon.

As winter deepens, snags of dead standing trees provide critical habitat for many avian residents. When hemlock or spruce die, they stand until toppled by wind and decay. Strong winds often snap off the tops, making these trees susceptible to heartwood-rotting fungi. Once the wood is softened by heartrot, avian excavators hollow out cavities that provide insulation from the cold, shelter from wind, rain, and snow, and protection from noctur-

< [97] **John Paige**
Harlequins
Watercolor, 35 x 25 cm

[98] Susan Ogle
Low Tide
Oil on canvas, 51 x 61 cm
"The oystercatchers were guarding their rocky nest, the eelgrass was exposed, and the clouds were stupendous. As the saying goes, 'All was right with the world.' The golden glow of the popweed on the upper reaches of the beach caused me to dig deeply into my oil paint box for an ancient, seldom-used tube of aureolin yellow."

[99] **Andrea Rich**
Alaskan Coast

nal predators such as great horned owls and pine martens. These cavities are used by many of the small resident birds – golden-crowned kinglets, chestnut-backed chickadees, pine siskins, brown creepers, redpolls, dark-eyed juncos, red-breasted nuthatches, and white-winged and red crossbills, among others. On stormy winter nights, with over eighteen hours of darkness, several species may huddle in a single hole. During brief daylight hours, foraging birds find dormant spiders and overwintering insects in bark crevices and decaying wood of tree snags.

During years of heavy snows, mountain goats move into lower elevations of the Chugach Mountains. They seek out steep, forested sites near cliffs, where they browse on hemlock and forage for available shrubs and grasses. On forested islands in Orca Inlet, Sitka black-tailed deer[7] move into lower elevations where less snow accumulates in the forest and where woody shrubs and forbs can still be found.

Other forest residents remain active during the winter. Voles, lemmings, and shrews feed on stored caches of grasses, roots, and bulbs. Squirrels reap their stores of seeds, berries, cones, and fungi. Spruce grouse make do with spruce needles, while snowshoe hares eat spruce needles and twigs as well as more nourishing willow and alder buds. These creatures fall prey to weasels, martens, mink, wolverine, coyotes, wolves, and owls.

For many of the forest residents, winter is a time of hanging on until the spring.

ALPINE TUNDRA AND ROCKY SLOPES

APRIL

When the hooligan return to the Delta in April, the high country of the Chugach and Ragged mountains is still locked in winter. The tundra, all alpine in this region and occurring mostly above 1,500 feet (460 m), is under snow from October to June.

> [100] **Dylan Lewis**
> *Cliff and Forest by the Bay*
> Mixed media, 27 x 37 cm

[101] **Andrew Haslen**
Belted Kingfisher, Katalla
Watercolor, 54 x 75 cm
"A bird I had not given much thought to was the belted kingfisher. The patterns it made in the landscape interested me, and I could have spent the whole trip making studies of flowers and kingfishers."

[102] **Andrew Haslen**
Belted Kingfisher over
Kattala Airstrip
Mixed media, 54 x 74 cm

[103] **David Bennett**
Steller Sea Lions and
Harlequin Ducks
Watercolor, 73 x 54 cm

On some of the south-facing slopes and wind-swept ridges, small bands of mountain goats paw for mosses, lichens, and low-lying shrubs.

MAY

The snow clinging to the steep, high slopes is not the crisp, white snow of winter. It starts to sag, developing a series of horizontal wrinkles under cornices and on avalanche slopes. Whole sheets pull loose with a whoosh! and rumble and thunder down slopes, leaving behind a rumpled, stained path.

Protected niches on south-facing slopes, which become snow-free relatively early, harbor the first flowers of spring. Alpine forget-me-nots crowd wet, rocky slopes at the edge of snow fields. Delicate grove sandwort, mat-forming arctic sandwort, and the evergreen prickly saxifrage bloom in dry, rocky areas.

The shrill whistles of hoary marmots ring out from talus slopes near treeline.[8] True hibernators, hoary marmots emerge from their dens in early May to find food and mates. When foraging for moss, lichen, grass, roots, and early flowering plants, members of the colony keep a lookout for coyotes, wolves, brown bears, and bald eagles.

By late May, the massive bird migration passing through the lower elevations finally spills over into the alpine tundra. A surprisingly high number of thrushes, warblers, sparrows, finches, and other passerines continue their journey westward through the snow-free tundra.

JUNE THROUGH JULY

As the snow pulls back to expose the rocky slopes and alpine tundra, mountain goat nannies give birth to single kids. The young are precocial and can scramble after their mothers on steep hill-sides within a few hours of birth. Females with kids and immature goats form nursery bands, while adult males hang out in bachelor groups. The goats graze on grasses, sedges, moss, lichens, and mat-

[104] **Bruce Pearson**
Sea Otter Study

Watercolor, 29 x 41 cm

Andrew Haslen: "Sea otters were another creature I had wanted to see, possibly a good view through the telescope.

In reality we boated into the middle of rafts of them, 100 or 200 at a time, almost able to touch them, swimming on their backs like little old men."

forming shrubs such as crowberry, moss and mountain-heather, kinnikinnick and alpine bearberry, round-leaf and arctic willow, and dwarf and bog blueberry.

Racing the short alpine summer, several birds brood clutches tucked in tundra micro-habitats by early June. Savannah sparrows nest in wet, grassy meadows, while rosy finch choose dry, grassy areas or even barren, rocky slopes. Wandering tattlers and water pipits nest in scrubby vegetation along rocky, mossy streams. Willow and rock ptarmigan cluck and strut, their nests hidden in shrubby areas.

A rich variety of wildflowers blooms above treeline. On dry, steep, rocky slopes, clumps of pink moss campion bloom with twinflower, alpine azalea, alpine arnica, eskimo potato, pixie-eye primrose, juneberry, yellow oxytrope, and fragrant woolly lousewort. On moist, rocky slopes, alpine veronica, few-flowered corydalis, frigid shooting-star, long-stemmed alpine heuchera, and mountain harebell grow in bright patches. Tundra meadows are a carpet of shrubs and flowers. Purple arctic lupine and Lapland diapensia bloom with western buttercups, roseroot, glaucous gentian, yellow paintbrush, pink plumes, alpine lily, spiraea, milk vetch, and meadow bistort. On warm, sunny days in early June, the subtle, sweet aroma wafting up from the tundra is unforgettable.

Tiny mammals scurry about the tundra. On well-drained hillsides, dusky shrews hunt insects, spiders, and small grubs in scattered patches of dwarf

willow, while singing (Alaska) voles forage in heather for grasses and seeds. Meadow voles and brown lemmings prefer grassy meadows, while masked shrews, northern redback voles, tundra voles, and northern bog lemmings dominate moist heaths and wet meadows. The *Microtus* voles (singing, meadow, and tundra) live in small colonies, and members help build a network of crisscrossing runways. Solitary northern redback voles use runways built by *Microtus* voles or the northern bog lemming.

Wolverines wander the remote, high areas of the Copper River Delta. These animals rely on areas with permanent snowpack to den, rear young, escape predators, and hide food. Tough and fierce, kits are weaned by midsummer and can forage for themselves by late fall. These solitary animals travel extensively in home ranges in search of voles, red squirrels, snowshoe hares, ptarmigan and other birds, and carrion.

AUGUST THROUGH SEPTEMBER

By mid-August, the steep slopes and high tundra glow with brilliant golds, reds, and greens as the deciduous shrubs respond to the chill of fall. Tucked in the matted tundra are tiny globes of purple, black, blue, and red, the fruits of juneberry, crowberry, bearberry, blueberry, kinnikinnick, and heather.

In September, the high peaks and upper ridges of the Chugach and Ragged Mountains and the Heney Range are often cloaked in peculiar, lenticular clouds. As these high areas chill, warm masses of air in the immediate vicinity condense and leave light dustings of snow, so-called termination dust, signaling the end of summer. It's enough to send the hoary marmots into their dens for the winter.

As fall storms build in intensity in the Gulf of Alaska and travel inland, more warm-air masses get hung up in the high country, billowing into great shrouds. Snowline, sharply etched on the dark slopes, creeps towards treeline.

< [105] **David Rosenthal**
Winter, Heney Range
Oil, 76 x 81 cm
A view along icy summit ridges out to Hawkins and Hinchinbrook Islands.

OCTOBER THROUGH MARCH

Usually by late October, the alpine tundra and rocky slopes are covered in snow that won't disappear until June. Despite the freezing conditions and harsh storms that buffet this region for over six months, many animals remain active throughout the winter. Mountain goats usually stay in steep forested slopes, but will climb slopes and ridges kept snow-free by strong winds. In the meadows, rodents scurry about under and on top of the snow as they feed off summer caches and harvest the seeds from grasses bent by the snow, while shrews eat plants or small rodents. These mammals in turn are staple food for coyotes, weasels, wolverines, and owls.

Still, life is relatively sparse in this region in winter, compared with the sheltered rain forests and coastal bays. And it's a long wait until spring.

[106] **Colin See-Paynton**
Trumpeter in Sun and Snow
Wood engraving, 18.5 x 18 cm
Some swans choose not to migrate south but overwinter at the mouth of Eyak Lake where it spills over into Eyak River, an area usually free of ice. As daylight lengthens, these overwintering swans are the first to initiate nests.

HUMAN ACTIVITIES

NATIVE CULTURES: WAR AND TRADE

Ever since the ice began retreating some ten thousand years ago, the Copper River Delta has served as a crossroads where different cultures have met in exploration, trade, war, and peace. The Delta region was the southeastern limit of the Pacific Coastal Eskimo culture (Chugach-Aleut),1 and it was the northernmost point to which the Tlingit of southeast Alaska ventured. Upriver were the Athapaskan Ahtnas, living on the interior Copper. The Chugach-Aleut, with their sealskin bidarkas, and the Tlingit, in open dugout cedar canoes, were seafaring people who pursued seals, sea lions, sea otters, fish, and other marine prey. The Chugach-Aleut established villages in the islands and fjords of Prince William Sound, to the west of the Copper

River Delta, and the Tlingit set up camps to the east of the Delta, along the Yakataga coast.

In between the Chugach-Aleut and the Tlingit were the Eyak, a smaller and less aggressive tribe than their powerful coastal neighbors. Related to the interior Athapaskans, the Eyak migrated over the icefields to the coast perhaps 3,000 years ago. Never a large group, the Eyak culture today is known from the few survivors of their last stronghold, the Copper River Delta.2 They lived, for the most part, off the bounty of the coastal waters, river, and land. Salmon were the chief food, taken from April through December and dried for the winter months. Waterfowl were captured during their autumn molt (when unable to fly). Grouse and ptarmigan were trapped or snared year round, as

<< [107] **Cordova harbor today**

(PHOTO: PAT AND ROSEMARIE KEOUGH)

< [108] **Fishing fleet, Copper River flats (circa 1940)**

(PHOTO: COURTESY OF CORDOVA HISTORICAL SOCIETY)

∨ [109] **Cordova, 1919**

Cordova in its early days as railroad terminus and seaport for the world's richest copper mine.

(PHOTO: COURTESY OF CORDOVA HISTORICAL SOCIETY)

Cordova, Alaska. "The Copper Gateway"
Scarborough
1919

were beaver, muskrat, mink, hare, and wolf. Mountain goats, and brown and black bear were often hunted, seal and sea otter slightly less so. Halibut and fat-rich hooligan (eulachon) were caught when available. Mussels and clams were gathered from the beaches, and berries, grasses, and other plants were harvested in season.

The Eyak people were pivotal in trade. Ahtna people from the Copper Basin brought copper as well as hides of moose, caribou, and other interior mammals to trade with the Eyak for sea goods from the region's coastal tribes. From the Ahtna (via the Eyak), the Tlingit people obtained their precious copper, which they worked into their art and culture by melting it with coal, also obtained through trade with the Eyak.

NATIVES, RUSSIANS, EUROPEANS, AND EARLY AMERICANS: TRADE AND EXPLORATION

During the mid-eighteenth century, Russian and European explorers crossed paths along the North Gulf coast of Alaska. In 1741 Vitus Bering, a Danish captain in service to the Russian czar, sailed the St. Peter from Siberia, first making landfall along the lee of Kayak Island. He dropped anchor offshore, and named the cape's ominous pinnacle of rock St. Elias, in honor of the saint of the day. Austrian naturalist Georg Steller was rowed ashore with a watering party and spent ten precious hours gathering plants, birds, and even some arrows, tools, and prepared fish from a Native camp, while the residents hid in the nearby forest. Steller's party left an iron tea kettle, pipes, beads, silk, and tobacco in exchange for the items they took.

So began a period of rapid change for the Native people, especially for the Chugach-Aleut, whose skill in hunting the sea otter was exploited by the Russians for their own trade. In 1788, seven years after the Copper River was first noted in journals of Russian explorers, the Russians built a small trading post on the Delta. By 1793 a large fortress, Fort Constantine, was established at Nuchek on Hinchin-

[111] *Alaska Natives with ten double bidarkas at the Martin Islands (circa 1920).*

(PHOTO: COURTESY OF CORDOVA HISTORICAL SOCIETY)

brook Island among the Chugach-Aleut people. Nuchek also became the site of the first Russian Orthodox church in this region. The Russians conducted a lively trade in sea otter pelts. (By 1830, sea otters were so rare that the Russians forced conservation measures on the trade.)

From Nuchek, the Russian hunters and traders explored the Copper River, but both the river rapids and the fierce Ahtna Natives repeatedly foiled attempts by newcomers to gain upriver passage. Finally the Russians established a fort near Taral at the confluence of the Copper and Chitina Rivers in 1819 which remained open until 1848, when the Ahtna wiped out a nearby exploration party. Neither the Ahtna nor the rapids, however, stopped the exchange of Russian and European goods or the introduction of their diseases. A smallpox epidemic swept along the Native trade routes in the 1830s, wiping out entire villages.

Alaska passed from Russian to American control in 1867 and some of the new owners were fascinated with the rich copper ore, and spearheads, arrowheads, utensils, and jewelry of pure copper being

< [110] *Paddling along the toe of Sheridan Glacier.*

(PHOTO: PAT AND ROSEMARIE KEOUGH)

< [112] **John Paige**
Dusky Canada Geese
Oil on board, 50 x 39.5 cm
Nearly the entire world popula-
tion of 11,000 dusky Canada
geese breeds in the Copper
River Delta.

traded down the river through the Eyak people. In 1884, with help from the Eyak and Ahtna Natives, U.S. Army Lieutenant William Abercrombie and his party succeeded, with great difficulty, in getting beyond the Childs and Miles Glaciers, only to be forced back by what became known as Abercrombie Rapids. One year later, in 1885, Lieutenant Henry Allen and his party completed exploration of the Copper River to its headwaters (and, from there, succeeded in crossing to the Yukon River and descending it to the Bering Sea).

[113] A photo of the Million Dollar Bridge where it crosses the Copper River just above Childs Glacier, visible in the background (circa 1930)

(PHOTO: COURTESY OF CORDOVA HISTORICAL SOCIETY)

EARLY AMERICAN SETTLEMENTS IN THE COPPER RIVER REGION

The Copper River Delta had captured the attention of frontiersmen for other reasons: salmon and oil. In 1887, the Odiak cannery was built near the old village of Eyak on the western edge of the Delta. Three other canneries, strung along the deep-water shore of Orca Inlet, were in business within five years. Then, on the eastern edge of the Delta, oil was found by early American settlers when a hunter literally slipped and fell into a pool of it near the Native village of Katalla (an Eyak word for oil). The first oil well in the Alaska territory was drilled in Katalla in 1902, and the oil rush was on.

But other discoveries upriver on the Copper had an even greater influence on the future. In the late 1890s, on the heels of the Yukon River gold rush, gold and then copper were located in the Copper Basin by prospectors who entered the territory through the seaport of Valdez in Prince William Sound. By the early 1900s, over 4,000 mineral claims were staked in a large area of the Copper and Chitina Rivers.

There was a driving need to build a railroad to service this copper-rich country from a suitable deep-water port: contending sites were Valdez, Katalla, and a new town called Cordova, which had been hastily staked on the Native village of Eyak in 1906. The Eyak cannery buildings, abandoned through a merger and consolidation of fishing companies into the Alaska Packers Association, were sold to legendary railroad builder Michael J. Heney ("Give me enough snoose [snuff] and dynamite and I'll build you a road to Hell!"). Heney turned the buildings into the terminal grounds and headquarters for the new railroad. By 1908 the Cordova townsite, which had been moved to the shores of Orca Inlet, was the undisputed terminus for the Copper River & Northwestern Railway. Valdez's contending bid was doomed when it appeared that Bering River coal, on the east side of the Delta and nearer to Cordova, would become available for railroad use (though the coal fields were subsequently closed to entry by President Theodore Roosevelt, to prevent corporate monopoly).3 Katalla's hopes literally washed away when a great ocean storm in November 1907 wiped out the dock and breakwater.

Cordova's transformation from Eyak village to bustling railroad terminus and seaport was rapid: in four years Heney and his men, financed by the Guggenheim-Morgan syndicate, completed the railroad on Native trade routes along the Chitina and

[114] *Fishing nets*

(PHOTO: PAT AND ROSEMARIE KEOUGH)

[115] *Brightly colored, small salt-barrels were once used to salt herring.*

(PHOTO: PAT AND ROSEMARIE KEOUGH)

Copper Rivers, crossing the latter between the spectacular Childs and Miles Glaciers with the legendary Million Dollar Bridge. The 196 mile (316 km) line to the mines near McCarthy is still considered one of the greatest engineering feats, under the most adverse conditions, in the history of Alaska.4

The copper era blazed through the region like fireworks – fiery, dramatic, and short-lived. For a while, the Kennecott Mine near McCarthy was the richest copper mine in the world (some of the ore was assayed at nearly 80% pure copper). When the railroad closed in 1938, owing to low copper prices, labor problems, and lack of access to local coal fields, the Kennecott had become the eleventh largest copper-producing mine in the world, a claim it still holds. In the wake of the closure, the railroad boomtowns McCarthy, Kennicott, and Chitina became ghost towns, and Katalla soon followed suit, unable to contend with the double loss of its oil production, due to a refinery fire in 1933, and then closure of its own local copper mines.

Also by the time the railroad closed, unchecked epidemics and diseases had wiped out over half of the Natives in the region, resulting in the loss of much traditional knowledge and thoroughly disrupting the social structure underpinning the millennia-old culture.

Cordova survived the closures of the copper mines and railroad with relative grace, because the town by then had ten canneries, the salmon harvest was at a record high and herring salteries were in full swing. During the 1920s and 1930s, the town gained renown as the 'Razor Clam Capital of the World.' (Pioneer Canneries had started commercially packing razor clams in 1916, after two years of trial and error.) In addition, a process for canning crab meat had been gradually developed and improved during the 1920s, and by the early 1930s, Cordova had become one of the leading exporters of crab in the Alaska territory.

"All the rest of Alaska's products put together can't equal the value in cold dollars and cents that her fisheries annually yield," boasted a 1940 pamphlet distributed in Cordova for the Second Annual Sea Food Festival. "From fish, and not precious metals, originate the taxes that build Alaska's roads, conduct her schools and execute her government. It's fish, not furs, that provides the bulk and the revenue for the Territory's transportation and commerce. Up from the sea, not the pockets of wealthy sportsmen, roll the dollars that support our mercantile trades, our service and secondary industries and our professional and technical men. Political and military spending may, for a time, assume great importance in the Territory's fiscal affairs; but, when their zenith has been reached and is waning, sea foods and allied products will still be riding in on the flood tides to furnish a stability needed by every populace."

Cordova was on a roll, and it was a good thing, because events in the Copper Basin once again played a heavy hand in the Delta's future. During the 1940s, for the first time since the gold rush, the Copper Basin experienced a development boom driven by World War II. Newcomers poured into the area to build roads linking the upper watershed with Anchorage (Glenn Highway), Valdez and Fairbanks (Richardson Highway), and the outside world (Alaska and Tok Highways). Airfields were built. This infrastructure and new technologies made the old ways of travel and trade obsolete for residents in the Copper Basin.

During this same period, there was a tremendous push to build a Copper River Highway on the old Copper River & Northwestern Railway right-of-way (which had been given to the federal government for this purpose by the Guggenheim-Morgan syndicate). A two-lane gravel road on the old railbed was constructed from Cordova across the west Delta and up to the Million Dollar Bridge, a distance of 50 miles (80 km). But progress for this rails-to-road transition from Cordova all the way to the Alaska Interior was slow, and it halted completely when the violent 1964 earthquake knocked

[116] *City airport, Cordova*
As in most parts of Alaska, airplanes have become essential to transportation in the Delta.
(PHOTO: PAT AND ROSEMARIE KEOUGH)

CORDOVA TODAY

Cordova is the only community on the Copper River Delta. The year-round population of 2,500 doubles in the summer with a seasonal influx of fishermen and cannery workers from other regions of Alaska and the Pacific Northwest. About twenty percent of the residents are Eyak, Chugach-Aleut, and Tlingit Natives, making Cordova the largest Native village in the Copper River–Prince William Sound region.

The social fabric of the community is a blend of diverse cultures and lifestyles. Natives, who are defined by their subsistence culture, also participate in the cash economy, while non-Natives, who rely primarily on the cash economy, also participate actively in the harvesting and sharing of natural resources. The town's main economy, commercial fishing, is an economic extension of a subsistence lifestyle. There are many parallels between the subsistence lifestyle and the commercial fisheries in terms of seasonal phases and social bonding. Often the first salmon of spring are shared by the fleet with the townspeople during a community salmon picnic.

Over ninety percent of the households in Cordova share with extended family and friends wild foods such as salmon, deer, moose, ducks, berries, jams, halibut, rockfish, shellfish, various plants, and miscellaneous treats such as marine invertebrates (chitons, snails, limpets, etc.) and mouse-nuts.5 Overall, residents harvest well over a dozen kinds of wild foods, averaging 400 pounds of wild food per household.6

The overwhelming majority of fish processed in Cordova are salmon (five species), which are caught in fisheries in the Copper River Delta and in Prince William Sound. Salmon fisheries are busy from mid-May through late September. Other important species include halibut, cod, and pollock. Herring was very important, prior to the oil spill, and will become so again, after their stocks recover to where there is a harvestable surplus (over the base

[1119] *Strips of salmon prepared for smoking*

(PHOTO: PAT AND ROSEMARIE KEOUGH)

[1118] *RJ unloading the salmon harvest*

(PHOTO: PAT AND ROSEMARIE KEOUGH)

< [1117] *Salmon on ice. Premium Copper River reds await processing.*

(PHOTO: PAT AND ROSEMARIE KEOUGH)

one of the spans of the Million Dollar Bridge off its piers. The river's shifting course over the next few decades ate away at the old railroad bed, pulling entire sections into the water and eroding the right-of-way.

Cordova, with no road to the outside world, has remained effectively cut off from the Copper Basin in commerce and in culture. Like the evolution of a species on a remote island, Cordova, in its relative isolation from mainstream twentieth-century society, has evolved a distinct culture and ambiance. Fishing has become a way of life for many residents, and the community's economy rises and falls with the fortunes of the fisheries, which are inextricably linked with the health of the environment.

Harvesting and Sharing in the Subsistence Culture

*"**Pre-gathering activities** include preparing the implements for gathering or capture, or giving lessons in using the implements, or a storytelling in the manner of folklore or of actual experience. The time spent together preparing for the activity is important: it prepares the individual and also strengthens bonds between co-gatherers and among the entire subsistence group.*

***Gathering activities** are of utmost importance, as they are the tests of lessons taught. Each activity helps measure the quality of time spent to demonstrate a skill, and to determine whether it is necessary to extend the time concentrating on the specific skill or to continue with the lessons of life. During the gathering cycle, individuals develop the skills necessary to earn and establish their success status.*

***Post-gathering activities** include food preparation, for immediate use or for a later time. Participants may work as a group, or several together, or as individuals. During the social gathering, events leading to the post-gathering are recounted and skills are reviewed.*

It is during these cycles of subsistence that bonding is strengthened and expanded. The sense of worth is solidified and new skills are learned. It is during these bonding times that our individual value is placed within our community, and we are able to understand what we must do to preserve our lives and to live in harmony."

Patience Anderson Faulkner
Chugach-Aleut Native, Cordova

level necessary to support other life in the Sound).

In Cordova, the harvest cycles turn with the seasons to become the calendar years that form a rural Alaskan lifestyle. All in the community who harvest from the land and sea are bonded through one central commonality: the sharing of natural resources. The economy, based on subsistence and fishing, creates an intricate web of social bonding, underpinning the community and linking people to place.

CORDOVA CALENDAR

APRIL-MAY
Herring return! Migratory people return as spring herring fisheries gear up. Bird watching out at Hartney Bay and Alaganik Slough. Cordova clean-up day. Fiddlehead gathering.
World-class bird watching with family, friends, and visitors. Shorebird Festival! Commercial salmon fisheries open on the Delta (first on west coast). Community salmon picnic. Potlucks galore. Avalanche season in the mountains.

JUNE-JULY
Commercial salmon fishing opens in the Sound. Hiking, camping, picnics, canoeing, kayaking, sailing, rafting, and surfing. Summer solstice celebration (longest day of the year, 19.5 hours, with long twilight). Family and friends visit from Outside. Strawberry picking in late July. Razor clam digging on outer beaches at very low tides.

AUGUST-SEPTEMBER
Deer hunting and silver salmon seasons open in August. Berry picking! (salmonberry, blueberry, cranberry, nagoonberry, currants) Stormy weather starts. Moose and migratory bird hunting seasons open in September. Many fishermen start to store fishing gear for winter. Mushroom picking.

< [120] **David Barker**
El Firma
Watercolor, 37 x 55 cm
"Now aptly named, this wooden fishing boat will never leave her final resting place on the mussel bank at the entrance to Heney Creek. A small spruce grows on her portside deck, and the flights of gulls will continue forever up and down Orca Inlet, at least while there are canneries at Cordova."

[121] *Artist Piet Klaasse on nets*

This 80-year-old master artist became a familiar and favorite face in Cordova's fish processing plants, fishing boats, and private homes as he skillfully depicted lifestyles in this community.

(PHOTO: PAT AND ROSEMARIE KEOUGH)

[122] Piet Klaasse
Bonnie Mending Nets
Mixed media, 28 x 19 cm

[123] **Piet Klaasse**
Salmon Delivery
Mixed media, 19 x 28 cm

OCTOBER–DECEMBER

Meeting season, potlucks, crafting, and music jamming begin in earnest. Bait-herring fisheries in November. Incredible ice-skating (and bonfires) at Sheridan Glacier, Eyak Lake, and on the Delta before snow flies. Back-country skiing, snow machining, and snowboarding. Downhill skiing on a good snow year (original single-seat chair lift imported from Mt. Baldy, Sun Valley, Idaho). 'Stage of the Tide' theater begins winter drama series. Winter solstice celebration (longest night, 19.5 hours, with short twilight). We start to gain light now!

JANUARY

Town meeting on major issue facing community. Cabin fever strikes time to go Outside on vacation. Pollock fishery opens.

FEBRUARY

Ice Worm[7] Festival to cheer us through winter. Plenty of races, food, and a parade! Fireworks! Meetings start to grow old.

MARCH

Spring is in the air. Skunk cabbage is sure sign of spring! Good spring skiing. Meetings are scheduled like crazy (what happened to winter?!). Dungeness crab and Pacific cod fisheries open.

[125] *Salmon are harvested commercially by two methods: gillnetting and seining. Here, the cork line of a seine net in Prince William Sound alters the pattern of wind and wave reflections: the net appears to be fishing for sunlight.*

(PHOTO: DAVID GRIMES)

‹ [124] **Piet Klaasse**
Gillnetter Bringing in the Set
Watercolor, 26 x 35 cm
The Copper River Delta salmon fishery is a mainstay of the local economy, particularly since the collapse of the herring and pink salmon fisheries in nearby Prince William Sound after the Exxon Valdez *oil spill.*

[126] Piet Klaasse

Marcene Mending Nets

Mixed media, 31 x 24 cm

[127] Piet Klaasse

Mark Krzympiec

Conté, 28 x 19 cm

[128] Piet Klaasse

Dee Dee, Can Sealer

Conté, 28 x 19 cm

"Ones culture should not be compromised or sacrificed for another cultures greed"
Jamachakih (Dune Lankard)
Eyak name: Little bird that screams really loud and won't shut up!

[129] **Piet Klaasse**
Nilatoa
Drypoint etching, 28 x 19 cm

[130] **Piet Klaasse**
Jamachakih
Conté, 37 x 28 cm

Cannery Row, Cordova, Alaska

[132] **Dylan Lewis**
Blue Boat
Watercolor
*Seiners such as this are used
to fish salmon in Prince
William Sound.*

< [131] **Piet Klaasse**
Cannery Row, Cordova
Watercolor, 35 x 53.5 cm

[134] **Dylan Lewis**
House at Cannery Row
Watercolor, 28 x 21 cm

[135] **Dylan Lewis**
Cabin Interior, Softuk
Watercolor, 26.5 x 30 cm
The U.S. Forest Service maintains a system of forest cabins throughout the Chugach National Forest.

< [133] **Dylan Lewis**
Spike Island and Cannery
Watercolor, 19 x 30 cm

[137] *Keith Brockie working in the old cannery netloft*

(PHOTO: PAT AND ROSEMARIE KEOUGH)

< [136] **David Barker**

Cannery

Pastel, 45 x 54 cm

"*It is an architecture born of necessity that reflects its purpose in a wonderful arrangement of iron and timber, all poised high on pilings above the tide-stream.*"

[138] **Keith Brockie**

Cannery Eagle Portrait

Watercolor, 35 x 25.5 cm

Studies of an adult bald eagle on a piling behind Cannery Row

139

< [139] *Logging operations on the Delta and throughout the region generate a high waste factor and target low-end commodity markets. Much of the 'trash' wood and the low-value wood sold for pulp has value – but in different markets. Pilot projects in certified sustainable forestry for private land owners are being encouraged in this region.*

(PHOTO: PAT AND ROSEMARIE KEOUGH)

Options for the Future

The Delta's rural economy is driven by harvest of natural resources (primarily fish, and recently, timber), although tourism is steadily increasing. Currently no comprehensive long-term plan exists for resource development or management in the Copper River watershed, largely because of political boundaries among different landowners. While fisheries resources haven been managed by the State of Alaska for sustained yield, forestry resources have not. Tourism growth is haphazard, responding mainly to pressure by urban and industrial interests rather than to long-term planning by residents.

Spurred by the devastating Exxon Valdez oil spill in 1989, residents are starting to work together to diversify the rural economy in ways that will sustain the natural resource base on which their communities depend. Ventures that maximize the value of resources while not increasing the harvest are being encouraged. These ventures may include pilot projects in certified sustainable forestry, rehabilitation and restoration of clear-cut lands, and a fisheries waste reduction plant.

[141] **Juan Varela Simó**
Old Mine, McKinley Lake
Watercolor, 25 x 18 cm
By the early 1900s over 4,000 mineral claims were staked in a large area of the Copper and Chitina Rivers. Most, like the 'Lucky Strike' pictured above, never panned out much profit.

141

[143] *Eyak Grave*

(PHOTO: PAT AND ROSEMARIE KEOUGH)

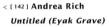

< [142] **Andrea Rich**
Untitled (Eyak Grave)
Woodcut, 25.5 x 33 cm
*"Near the edge of the forest
we came upon an old, tilting
fence. It had been erected to
surround the grave of some
person long dead. A large
tree grows out of the space
now and the cross with no
name is all that remains to
remind us that we all will be
gone soon. What will remind
others that we were once
here?"*

[144] **Siegfried Woldhek**
Steller's Jay in Clearcut
Watercolor, 33.5 x 49.5 cm
*Large scale clear-cutting is a poor harvest method on the Delta,
because the steep gradients, heavy precipitation, and frequent
earthquakes can result in high erosion and siltation of salmon
streams. Also, the patchy forest cover requires many access
roads, erosion from road-building activities has been found to
damage salmon and wildlife habitat in the Pacific Northwest.*

< [145] **Tim Shields**
North Arrow
Watercolor, 50 x 70 cm
A unique experience awaits visitors who make the extra effort to get to Cordova and take the extra time to stay and explore the community and the Delta. As a quiet community, a little off the beaten track, Cordova remains a pocket of the real Alaska, an unspoiled coastal fishing community with a rich history and stunning surroundings.

[146] **Tim Shields**
Shorebird Festival
Watercolor, 34 x 52 cm
Every year in early May Cordova hosts the Shorebird Festival to celebrate spring bird migrations.

APPENDIX A. THE *EXXON VALDEZ* OIL SPILL

On March 24, 1989, the *Exxon Valdez* grounded in Prince William Sound, spilling (as reported by Exxon) 11 million gallons (200,000 barrels) of oil. The oil eventually spread over 10,000 square miles (26,000 km^2) and oiled over 1,500 miles (2,500 km) of Alaska's coastline, killing more wildlife than any other spill in history. This spill triggered a series of events from which the Sound is still recovering. Genetic damage, reproductive impairment, and ripple effects detonated ecosystemwide injury and delayed recovery in a variety of invertebrates, fish, birds, and marine mammals.[1]

The oil spill devastated Cordova, economically and spiritually. Well before the economic damages became apparent, townspeople experienced intense anguish over the suffering and dying animals in the Sound, a gut response that doesn't go away for months, reflecting the deep spiritual, emotional, and physical connections people share with their natural environment. Disruptions to the harvest of natural resources further eroded the social and economic fabric of the community. The pattern of harvesting and sharing which forms the critical link between the community and the environment, and among families and community members, was shattered. Violent crime, domestic abuse, and calls to the mental-health-crisis hotline increased.

Economic damages piled on top of an already emotionally frayed community. In 1988, the year before the oil spill, Cordova was a bustling seaport, the nation's eighth largest, based on the dockside value of the seafood harvested ($46.4 million). Just four years later (1993), when the herring and pink salmon populations crashed, Cordova had slid to fifty-first place in a downward economic spiral that sent the community reeling from the loss of over thirty million dollars in annual income for its seafood harvest (compared with 1988). Unable to pay bank loans, fishermen lost boats, even fishing careers, and families lost homes that had been collateralized for boats and permits. The city's operating budget shriveled as its prime revenue base, the fisheries, dwindled in value and volume. Unable to find secure financial footing, many fishing families were forced to leave town. After hitting the 1993 low point, Cordova fisheries began a slow rebound as Prince William Sound salmon and herring stocks began to recover from the *Exxon Valdez* oil spill. By 1996, Cordova ranked twenty-second in value nationally ($27.8 million), still well below the pre-spill years.

Cordova's economic crisis could not be ignored. As the main fishing community in Prince William Sound, with long-term, spill-related injury appearing in the Sound's fisheries resources, Cordova had become the economic epicenter of the spill and was clearly in for a bleak near-term economic future. Local government held a series of town meetings during the winter of 1994 and 1995 to determine the community's desires. Residents agreed that efforts to diversify the economy should also maintain a healthy ecosystem to protect the fisheries (such as the Copper River) and protect the community's values, balancing economic and lifestyle needs.

But while Cordova residents had turned their energies to dealing with the aftermath of the *Exxon Valdez* oil spill, others had made plans for the future of the region.

WHO PLANS OUR FUTURE?

In 1991, using 'highway maintenance' funds, the State of Alaska's Department of Transportation started bulldozing along the overgrown and abandoned Copper River railroad bed to link the ends of the existing roads south from Chitina and north from Cordova, and to create a major highway for resource extraction and large-volume tourism. By its action, the state rekindled a long-standing debate pre-dating statehood in 1959.[2] If built, the Copper River Highway would connect the community to the state's road transportation network, irrevocably altering the town's unique character which has evolved, in large part, because of the community's relative isolation. The state's latest attempt to construct the Copper River Highway was stopped in court (the state had refused to obtain any of the federal permits necessary to begin the roadwork).

The issue, however, is far from over. Interest in building along the Copper River railroad right-of-way has revived because the State of Alaska is trying to draw visitors to the Copper Basin.[3] The State recently released its plan to construct a 'world-class', $27-million, 'nonmotorized' trail (instead of a road), along the Copper River between Chitina and Cordova. The proposal would, among other things, drive a core trail of 66 miles (106 km) through virtual wilderness, and would probably necessitate motorized access along the river, currently a premier rafting route, to service campgrounds and cabins for trail users. Many residents, landowners, and conservationists are skeptical of this plan, concerned that a trail would inevitably become a road over time. Others, who want a road, are equally concerned that the proposed trail would benefit only an elite group of ecotourists and would create a physical and political barrier to construction of a future highway. Some say there already is a world-class trail down the valley, the Copper River itself. In any case, significant increases in noise or traffic down the river would have enormous effects on the area's wilderness character, wildlife, and fisheries. Meanwhile, despite the lack of public consensus, and perhaps even aided by this lack of unity, the state ploughs doggedly on with its own plans.

The State of Alaska's Department of Transportation and Public Works is preparing a plan focused on regional transportation infrastructure needs and addressing, as its major issue, the best way to 'improve access to Cordova.' Transportation options to be examined include the Copper River Highway and improved ferry and air service.

During this same time period, the State of Alaska, City of Cordova, Eyak Corporation, and Chugach Alaska Corporation all expressed a keen desire to construct a road to Shepard Point, 6 miles (10 km) north of Cordova, and to build a deep-water port (closest available point to Cordova). The state and city claim the port's primary purpose is as a staging area for future oil-spill response and as a deep-water berth to enable larger cruise ships to visit the area. The Native corporations foresee using the port to export more timber and perhaps coal resources.

Alaska is one of only a handful of coastal states which allows (and actually promotes) oil and gas leasing in its coastal waters. (The temptations are great, since eighty-five percent of the state's operating budget currently comes from oil revenues.) Oil and gas lease sales are offered on a five-year cycle, depending on oil industry interest, by either the state or federal government. In the early 1990s, Arco Alaska, Inc., expressed interest in exploring for oil along the Cape Yakataga coast and westward to include part of the Copper River Delta. The Delta is extremely vulnerable to oil spilled along the Yakataga coast, because a strong westerly shore current sweeps the North Gulf coast. The state eventually lost its momentum for this lease sale when Arco Alaska, Inc., withdrew from negotiations.[4] However, Arco approached the state again in 1997, indicating continued interest in the area, which may reopen the issue.

Recent clear-cutting of rain forests started on Eyak Corporation lands in the Delta in 1988, the winter before the oil spill. Over the next decade, about 7,500 acres (3,000 ha) of Eyak Corporation's lands were logged near Sheridan Lake, Eyak River, and at Mile 17 ('haystacks'). During this time, the corporation also clear-cut about 1,000 acres (2,600 ha) in Orca Bay and Orca Inlet. Responding to environmental and aesthetic concerns from shareholders and local residents, the community-based corporation selectively harvested (with helicopters) in another 1,000 acres (2,600 ha) of land around Eyak Lake and River, and in Orca Inlet. Logging operations temporarily ceased in 1997 due to a drop in the cyclic market for timber commodities (whole logs and pulp). Eyak Corporation has entitlement to a total of 149,000 acres (60,300 ha) of land, primarily coastal forest, in the Delta and Prince William Sound.

Chugach Alaska Corporation, a regional Native corporation based in Anchorage, owns 73,000 acres (30,000 ha) of timber and some of the subsurface coal resources near the Bering River on the east side of the Delta. Eighty years after Theodore Roosevelt's epic effort to conserve this region, Chugach Alaska Corporation announced plans to develop its inholding in this area. The corporation stated that it plans to clear-cut the marketable timber on the land, about 8,000 acres (3,300 ha) starting in year 2000, then to mine the coal fields, and ultimately to use the area for tourism. These efforts would require construction of about 70 miles (113 km) of service roads in addition to a 25-mile (40 km) access road, which would cross over 200 salmon streams after branching off the existing Copper River Highway from Cordova at 42 Mile. This land east of the Copper River is currently a roadless wilderness. Chugach Alaska Corporation plans to start building the access road (a three-year project) during the summer of 1998, unless it is presented with more lucrative options for its land, such as conservation easements.

Adding to the drama, the U.S. government *Exxon Valdez* oil-spill restoration efforts (in addition to conducting scientific studies) have focused on habitat protection through the purchase of land or conservation easements from Native corporation landowners in the region affected by the spill. While habitat protection is essential for restoration, the current efforts by the government heavily favor outright land purchase over conservation easements. The prospect of losing forever the ownership of ancestral lands has created deep divisions among the Native owners. The offers are viewed as thinly disguised government land grabs, yet the promise of windfall cash distributions makes the decision very difficult. Natives have expressed keen interest in exploring alternate business and conservation proposals, which offer economic options other than clear-cutting or selling their corporation's land.

These major resource development and access improvements, some underway, some still under consideration, would irrevocably alter the character of the Copper River. A key problem is that individual proposals are being considered and decisions are being made without sufficient attention to the complex overall health of the region, without any questioning as to cumulative effects, and with minimum consideration of the wishes of affected residents. Nowhere in the calculation of these projects is there a real look at sustainable alternatives, or a consideration of what is the true, real wealth of this region.

[147] PHOTO: COURTESY OF CORDOVA HISTORICAL SOCIETY

A Natural Alliance

"I see disturbing parallel universes of Native and non Native. Nowhere in the region today will you find a place where people can communicate and break down barriers. We need this more than anything else. The time is well past that the two parallel universes need to communicate and talk. Everybody's truth needs to be listened to. If you don't get Natives involved, then you have an unnatural alliance and you will get faulty decisions."

– Wilson Justin
Ahtna Native, Chistochina

In September 1997, after a year and a half of meetings, discussion, and planning, residents participating in the Copper River Watershed Project voted to incorporate as an independent, nonprofit organization. Honoring equal and diverse representation from the Copper Basin and Delta, the residents from the two regions chose a founding board of Native directors. In the Copper River Delta, a steering committee was formed and work groups for forestry, fisheries, and tourism projects were set up; corresponding groups are anticipated for the Copper Basin.

As a coalition of residents from the twenty-two communities in this 26,500 square mile (70,000 km^2) Watershed Project's region, the first step is to define collective priorities for community values and future development. With these priorities in mind, the Project will seek to diversify the economy while maintaining the region's resource base and sustaining its cultural heritage. This project is unique in Alaska, and possibly among other such efforts, for its emphasis on economic, rather than regulatory, incentives and its holistic approach to sustainable communities.

The notion of 'sustainable development' was presented at a March 1996 community workshop as an approach to economic development which accounted for all of a community's assets – economic, natural resource, and cultural/social – when planning for growth. From this workshop and the four that followed, residents recommended focusing on forestry, fisheries, and tourism as the region's business sectors considered to be strongest and most likely to grow. Money was raised for a regional resource assessment that would identify previously untapped resources, potential markets, and infrastructure needs. The resource team was charged with forming economic links among the three resources to reveal how development of one resource could benefit users in other resource areas.

At the same time, to make sure that business projects being proposed would meet residents' priorities, residents were asked to express what they valued most about this region. Three themes sounded consistently: healthy fisheries, access to wildlife for viewing

and hunting, and a strong community. For Cordovans, this latter meant a "small town in a wilderness setting," "off the beaten track," with a diverse population full of talents, expertise, and skills, and with safe streets and safe schools. For the future, people stated that they would like to see a diversified economic base to stabilize the seasonal economic swings, healthy fish and wildlife populations, locally owned and operated businesses, and retention of a small-town atmosphere.

Economic indicators have often been used as primary measures of a community's well-being (average income, property values), but they clearly do not encompass all of the virtues residents might be seeking. So the next step in the Watershed Project was to draw up a list of indicators in three categories – natural resource, cultural/social, and economic – for tracking progress in growth and community health, and to ensure that community core values would not be diminished or lost as the economy is diversified.

Many indicators fall mainly into only one of these categories. Monitoring the strength and harvest of annual salmon runs, tracking game populations (e.g., waterfowl, deer and moose), monitoring regeneration of trees in a clear-cut or the use levels of recreational facilities – to detect warning signs of stress and overuse – are examples of indicators for the health of natural resources. Employment distribution by industry, and seasonal wage and salary employment trends are examples of economic indicators, while volunteer participation in the community is a social indicator. Cordovans boast of a high level of volunteer activism, which is important for strengthening social bonds and expressing a commitment to youth, neighbors, and friends.

Some indicators overlap and address more than one of these categories. When looking at the number of households that harvest subsistence resources (salmon and other fish, deer and moose, berries and other plants), the data reveal important information about non-cash supplements to the household economy. Indicators also reflect at what level people participate in an activity that is regarded as highly social: residents hunt game and pick berries in groups, and over ninety percent of those who harvest wild foods share their bounty with family and friends.

Taken together, these indicators can be used by Cordova and the up-river communities to prepare for the growth of tourism and other economic shifts by regularly taking stock of their assets. By identifying and monitoring indicators for community core values, community leaders choose to make the economy work for their particular community. By recognizing that economic decisions bear long-term social and environmental costs and benefits, business leaders become aware of the real impact of each decision. By working to increase the benefits while reducing the costs, all residents choose to make their communities sustainable.

FORESTRY OPTIONS

We need new ways of looking at an old forest: how can landowners realize greater economic, social, and environmental benefits from their home forests and keep them healthy? New approaches to forest conservation and to wood-product marketing can provide answers and solutions.

Most of the high-value timber in the Copper River watershed was selected by Native corporations as part of the Alaska Native Settlement Claims Act (1971). A significant problem for landowners throughout the Copper River watershed is that logging operations currently generate little net economic gain. Subarctic conditions limit forest productivity and affect wood quality.

Sitka spruce on the Delta is at the northern end of its range, growing in narrow bands of greenery sandwiched along the mountainous coast between the ocean and the ice: it can take 100 years for a tree to grow to 7 inches (18 cm) diameter, and rot and fungal infestations are prevalent. Meantime, spruce bark beetles infest much of the interior white spruce forests.[1] Faced with (perceived) low-value wood and high transportation costs, most landowners resort to large-scale clear-cutting in an attempt to maximize profits. A few short-term jobs are created, but profit margins are very low and fleeting. Native landowners typically receive little compensation in the way of dividends for liquidating vast tracts of forest that may not be restored for generations. Conservation deals, however, recognize that ancient forests or wetlands are already highly developed systems, producing a rich bounty in perpetuity if conserved.

The forest research conducted by the Copper River Watershed Project found new forest products and more profitable markets for existing products. The resource team observed that logging operations throughout the region generate a high waste factor and target low-end commodity markets (round log and pulp export) rather than seek out value-added markets. Much of the 'trash' wood and the low-value wood sold for pulp has market value in different markets. For example, landowners in Canada sold their trash logs (worth $0.25 per cubic meter as trash) to the log-home industry and received $80 per cubic meter. In another instance, 'characterwood' also has its niche markets. Traditionally priced down as defective, wood characterized by knots, wormhole- and insect-tunneling, mineral streak or stain, small pitch, gum and bark pockets, and grain and color variations can be worked into log homes and accessories, furniture, and casegoods at a dramatic increase in profit. Based on direct buyer interviews with custom log-home producers, some producers receive a thirty-percent premium using 'character logs.'

There also exists an established and rapidly growing market for over twenty non-wood forest products found in the Copper River watershed. For example, the root bark of devil's club, the spiny bane of hikers, is used as a blood-sugar regulator (insulin substitute) for diabetics, and has a market price of $22 per pound dry weight. Many potential non-wood forest products are abundant in the watershed and are currently little used, such as lady fern for the dried floral market, Sitka (diamond) willow for crafting fine walking sticks, and spike-rush and sedges for basketry.

Harvesting and local processing of non-wood forest products can create jobs without the degree of social and environmental disruption associated with large-scale clear-cuts. In Oregon's Willamette valley region, developing such products decreased the volume of wood products harvested from the forest by one-third, while increasing overall profits of landowners by three-hundred percent. This approach also benefits local tourism by creating place-specific local products to sell to visitors, while protecting the visual quality and ecological integrity of the lands travelers want to use.

Research found that the key to accessing maximum value for forestry products is through the International Forestry Stewardship Council's certified sustainable forestry program.[2] This is a voluntary program that uses market-driven incentives to achieve sustainable forestry management. The next step in this research will be to work closely with buyers and watershed public and private timberland owners to better understand local production capabilities and transportation costs, and to set out specific marketing and development plans for these, and potentially other, landowners.

FISHERIES OPTIONS

Like timber, most fishery products are only processed to a primary stage, which allows the most rapid shipment out of Alaska. Most salmon are headed-and-gutted, then either frozen or canned. Not only does this generate an enormous amount of waste, which is simply ground up and discharged into the ocean, but it also exports jobs and profits with its primary-processed fish. Most secondary, or 'value-added,' processing takes place outside Alaska.

There is growing interest in secondary processing of both fish and fish waste. Over the past few years, a small cottage industry, as well as the larger companies, has produced increasing amounts of varied salmon products, such as canned, lightly smoked, hard-smoked salmon jerky, salmon caviar, and skinless/boneless pink salmon fillet blocks.

However, with the exception of four large primary-processing plants, the community lacks substantial manufacturing facilities. The reasons for this include: (1) the high cost of diesel-generated electric power, compared with the Puget Sound area (24.9 cents per kilowatt hour compared with 6.2 cents/kW, respectively); (2) the high cost of labor in rural Alaska, compared with the Puget Sound area; and (3) the high cost of transporting raw materials to Cordova and the finished products out of Cordova.

The resource team found a spectrum of secondary-processed seafood products that might feasibly be produced in Cordova, such as retail-ready salmon portions, food-service-ready packaging, and seafood entrees. Also, the Watershed Project found ways of marketing these and other products, such as private-label packing for large mail-order companies; and a variety of waste-based products such as fish feed, poultry feed, fertilizer, pet foods and treats, and vitamin supplements (omega-3 fatty acids from fish oil, calcium from bone meal, etc.). Finally, the Watershed Project is working with Cordova's fishing industry to maximize the value of all seafood products produced in Cordova.

With the support of local banks and individuals, the Watershed Project is planning to serve as an 'incubator' to help existing and start-up businesses produce and market these secondary-processed and waste-based products. The goal is to increase the value of Cordova's fisheries-based economy, with no additional harvest pressure on the fisheries resources.

TOURISM OPTIONS

Cordova is off the beaten path, but it is strategically located in Alaska's most intensely used southcentral region.[3] Cordova is a gateway to four distinct and attractive natural destinations – the Delta, the Copper River, the Sound, and the Chugach Mountains. While it does serve as destination for some fishing and sightseeing and as host of the Shorebird Festival and Ice Worm Festival, the town has not marketed itself as a tourism destination.[4] Tourism has the potential to help diversify the local economy, but only as long as locally owned businesses are supported, year-round tourism is promoted to even out the seasonal swings, and the town and its surrounding natural attractions are protected from overuse.

Since 1990, Cordova has hosted about 10,000 visitors annually with a slow trend toward increasing numbers. This is significantly below visitation in other Alaskan coastal towns, where visitors number in the hundreds of thousands. In the past, these tourists arrived by ferry, plane, or private boat (a handful of small cruise ships visits Cordova).

In the summer of 1997, the City of Cordova announced that a major cruise ship line planned to include the town as a port of call, starting in 1998. This caused great consternation among many residents, who, once again, found themselves reacting to far-reaching plans and decisions that could alter the character of the community, yet which had been made with little public involvement. Initially, many residents were skeptical ("I haven't decided yet whether I'm going to build a gift shop or a fence!"), but this skepticism was slowly overcome by the city's willingness to work with the public and with key organizations to address residents' concerns and to prepare the community for the cruise ship visitors.

Joining efforts in a public process, the City of Cordova, the Cordova Chamber of Commerce, the U.S. Forest Service, Eyak Corporation, the Copper River Watershed Project, and interested residents initiated a local comprehensive planning process.

Through this process it has become clear that participants want a strong voice in guiding the type, magnitude, and location of tourism in order to maintain a healthy, productive natural environment and the quality of community life (or town character). Residents also want to avoid overcapitalization by ensuring that the desired level of growth is carefully matched to infrastructure capacity, and that tourism, not just the local residents, helps pay for increased infrastructure needs and other tourism-related services.

Using other communities as examples of a range of tourism growth, from small to large scale, residents have recognized the need to monitor change related to goals to avoid crossing the threshold of 'too much' tourism growth and resulting loss of quality of life. Monitoring social, environmental, and economic change is planned as an ongoing partnership amongst the city, chamber, and Watershed Project, based on the core values discussed earlier.

The underlying strategy behind all these tourism plans and projects is to base Cordova's tourism future around protecting community and environmental quality, so that there are real economic incentives for managing the amount and type of tourism growth.

The Watershed Project produced a report, *'Making the Most of Copper River Resources,'* which integrates the initial findings of the research team. Then, in concert with local government and leadership entities, Native corporations, and other businesses within the watershed, the Project is developing a working plan of actions to foster economic development and diversification compatible with achieving the goal of sustainable communities.

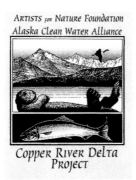

Logo design: Susan Ogle

NOTES

Chapter 1. Overview

1. In 1980 the Alaska National Interests Land Conservation Act (ANILCA) set aside 104 million acres of Alaska as federal land. The Copper River Delta was to be designated as a National Wildlife Refuge, but, as part of a compromise, it was designated instead as a National Forest to be managed by the Forest Service as a wildlife refuge. The Forest Service has upheld its mandate: to this date, the agency has not allowed logging on public land in the Delta. Logging, however, has recently taken place on privately held Delta lands.

2. The legal question of Alaska land ownership had to be settled in order to establish right of way for the Trans-Alaska Pipeline, built during the 1970s. The Alaska Native Claims Settlement Act of 1971 extinguished virtually all aboriginal claims to Alaska lands and in exchange created 13 regional Native corporations and 220 village Native corporations which were deeded title to nearly 44 million acres (18 million hectares) of land and given $962 million as start-up capital. Regional corporations selected lands for prime timber and oil value. The act guaranteed corporations access to their land selections. The act was supposed to provide security (land, jobs, improved living conditions, protection of subsistence resources) for Alaska's Natives, but by almost everyone's account, it has not worked. This act continues to play a huge role in shaping Alaska's environment and economy. Interested readers are referred to *Village Journey* by Thomas Berger, and *Sold American* by Donald Craig Mitchell.

3. Interested readers are referred to *Lost Landscapes and Failed Economies* by Michael Power for economic arguments that 'the quality of the natural landscape is an essential part of a community's permanent economic base and should not be sacrificed in short-term efforts to maintain employment levels in industries that are ultimately not sustainable' (dust jacket).

4. The *Exxon Valdez* oil spill is the most studied spill in history. In October 1991, the U.S. District Court approved a civil settlement and restoration fund plan in which Exxon agreed to pay $900 million over a ten-year period to reimburse the state and federal governments for clean-up costs and for restoration. The *Exxon Valdez* Oil Spill (EVOS) Trustee Council was formed under the civil settlement to oversee the fund. Among other things, the EVOS Trustee Council has sponsored annual studies to document spill-related injury and recovery. Studies are public and a summary is published annually. For more information about the EVOS Trustee Council, contact the Oil Spill Public Information Center, 3150 C Street, Anchorage, AK 99503, toll-free 800-283-7745 outside Alaska or 800-478-7745 within Alaska, or visit the web site at http://www.alaska.net/~ospic. The effects of this spill on the Prince William Sound ecosystem and Cordova are summarized briefly in chapter 4 and Appendix A.

5. After the *Exxon Valdez* oil spill, Cordova became a case study for leading rural sociologists studying the effects of a technological disaster in a natural-resource-based community. Interested readers are referred to *Exxon Valdez Disaster*, edited by Picou et al., for a nontechnical report of findings.

6. Founded in 1990, the Artists for Nature Foundation (ANF) has completed projects in the sensitive Dutch barrier island of Schiermonnikoog (*Wind, Wad, en Waterverf*, 1992), the Biebrza and Narew Valleys of northeastern Poland (*Portrait of a Living Marsh*, 1993), the *dehesas* and steppes of Extremadura in southwestern Spain (*Artists for Nature in Extremadura*, 1995), the Loire River and estuary in France (*That the Loire May Live*, 1996), the coastline of South Wexford in southeast Ireland (no publication), and a tiger project in the Bandhavgarh National Park in Madhya Pradesh, India (publication in preparation). For more information, contact ANF at Striepeweg 5, 7675 TG Bruinehaar, the Netherlands, telephone 31-546-681-253, or fax 31-546-681-765.

7. This Oral History has been handed down through five generations of Paula Underwood's family to be shared now "as a gift for all Earth's children with Listening Ears." Paula has established the Learning Way Company to enable further sharing. *The Walking People: A Native American Oral History* provides the basis for the retreats given by the Learning Way Company which focus on learning one's way through life, through the chaos of constant change. Information and books can be obtained by calling 1-800-995-3320. Web site explorations are welcomed at http://members.aol.com/totpress/.

Chapter 3. The Living Layer: The Ecosystem of the Copper River Delta

1. Some years the hooligan do not return in significant numbers, or even at all! Also, in response to large-scale climatic influences such as El Niño or extremely harsh winter conditions in the Pacific Northwest, arrival of other spring migrants may vary by several weeks, either earlier or later.

2. There are at least three species of cottongrass in the Delta (*Eriophorum augustifolium, E. russeoleum,* and *E. Scheuchzeri*).

3. Coastal temperate rain forests comprise a 3.2-million acre portion of the world's total acreage (74 to 100 million acres) of temperate rain forests. (For every acre of coastal temperate rain forest, there are roughly 90 acres of tropical rain forest currently standing.)

Coastal temperate rain forests play a critical role in cycling nutrients between the land and the sea: these forests produce more biomass (wood, needles, plant litter, moss, and organic soil) than any other forest type including tropical forests (200 to 800 metric tons per acre). Some of these nutrients are flushed to the sea and thus enhance productivity of nearshore coastal regions, which in turn support one of North America's richest fisheries.

4. The current population of about 1,000 moose on the Delta are all descendants of twenty-one calves flown in from the Kenai Peninsula and Matanuska-Susitna valley between 1949-58 by legendary bush pilot 'Mudhole' Smith. Prior to this time there were no moose on the Delta, because the Childs and Miles Glaciers blocked natural immigration down the river corridor. Smith worked with the Isaac Walton League and many local residents (Hollis Henrichs, Vina Young, and Ed King, among others), in successfully transplanting the moose to the lower Delta.

5. The Cordova Ranger District of the Chugach National Forest maintains a system of trails and Forest Service cabins throughout the Delta. Interested hikers are referred to *Take a Hike!* by the U.S. Department of Agriculture, Chugach Ranger District.

6. There are hundreds of different kinds of lichens in the Copper River Delta. Lichens are a type of fungi that have evolutionarily developed a symbiotic relationship with algae. The algae provides photosynthesized sugars to the fungus, while the fungus provides nutrients and shelter for the algae to grow. Lichens can be important food sources for browsing animals during the winter.

7. Between 1916 and 1923, nineteen Sitka black-tailed deer were transplanted to Hawkins Island in Orca Inlet and Hinchinbrook Island in Prince William Sound. This effort, undertaken by the Cordova Chamber of Commerce, was the first big-game transplant in the territory and is still considered to be the most successful.

8. There are some marmot colonies below treeline on the Delta, such as in the rocky talus slope near the Copper River at Mile 27.

Chapter 4. Human Activities

1. The Chugach-Aleut, who inhabited Prince William Sound (just west of the Delta), are part of the Alutiiq culture that ranges westward to include Kodiak and the Alaska Peninsula. The Alutiiq language, *Sugcestun*, is part of the Aleut-Eskimo linguistic family and most closely related to Yupik.

2. The Eyaks were the last Indian nation re-recognized in North America, in 1930 by anthropologist Frederica de Laguna. Today, Eyak Chief Marie Smith Jones, in her eighties, is the last Native speaker of Eyak.

3. "In 1907 President Teddy Roosevelt, squaring off against the J. P. Morgan and Guggenheim syndicate building the Copper River & Northwestern Railway to the copper mines, removed the half-billion-ton coal field (and others in Alaska) from public entry to prevent corporate monopoly. The same year, with his Chief Forester Gifford Pinchot, Roosevelt created the Chugach National Forest to more permanently protect the forest and wildlife resources of the Delta and Sound. These extraordinary and visionary conservation measures were of national significance. Pinchot, father of the Forest Service, was fired in 1910 by President William Howard Taft, after Pinchot blew the whistle on the syndicate's attempt to reclaim the Bering coal fields in a backdoor deal with Taft's Secretary of the Interior Richard Ballinger. The ensuing scandal rocked the nation and convinced Roosevelt, in defense of Alaska conservation, to come out of retirement and run against Taft in the next presidential election. The Republican Party was splintered, Roosevelt formed the independent Bull Moose Party, and Democrat Woodrow Wilson was elected. But the Copper River Delta remained protected until recent times." (David Grimes, from "First Do No More Harm," in *Alaska's Wild Voices*, Autumn 1997.)

4. The fascinating story of constructing the Copper River & Northwestern Railway is captured in *The Copper Spike*, by Lone E. Janson. Janson has many relatives in Cordova.

5. *Mousenuts* are the roots and grasses of cotton-grass, bulrush,

and other sedges, harvested by voles and lemmings and stored underground in food caches. Natives locate the caches and, traditionally, take only half the stored goods, or replace the sedge stems with other mouse food. Mousenuts are considered a delicacy among Alaska Natives.

6.   These data reflect harvests prior to the *Exxon Valdez* oil spill, according to Cherrington (1993) and Dyer et al. (1992). After the oil spill, subsistence harvests of wild foods declined up to 77 percent in some of the 15 natural-resource-based communities studied (including Cordova), while diversity of foods harvested declined by up to half, because of concerns about oil contamination (Fall 1990, 1991). In Cordova, recovery of subsistence harvests has occurred to some extent, but still lags behind the pre-spill harvests. This reflects the lagging recovery of the Prince William Sound ecosystem. It is generally believed that the subsistence harvests of people in Cordova (and other natural-resource-based communities in the Sound) will recover only when the Sound is fully recovered from the oil spill.

7.   Iceworms are segmented worms, related to earthworms and leeches. These dark, one-inch (2 to 3 cm) worms live in glaciers and adjacent perennial snowfields all year, surviving at near-freezing temperatures. They feed on snow algae and pollen grains, burrowing deep into the snow to avoid sun during the summer or freezing temperatures during the winter.

Appendix A: The *Exxon Valdez* Oil Spill

1.   According to studies conducted through the *Exxon Valdez* Oil Spill Trustee Council, reproductive impairment and genetic damage was found first in herring, then pink salmon, and is suspected in a number of other fish such as Dolly Varden and cutthroat trout. In 1992 and 1993, systemic problems appeared in the Sound when the pink salmon runs failed. Significantly, the back-to-back failure of these runs indicated that the adult pink salmon, exposed to the spill in 1989 as fry or eggs, failed to produce viable offspring. As the strong survive, the odd- and even-year classes of pink salmon are slowly recovering.

In 1993 the spring herring run crashed — almost 100,000 tons of herring had vanished. Of the remaining 20,000 tons, over one-third were infected with a virus that left visible lesions, and others were lethargic. Scientists speculated that toxic exposure to oil in 1989 had weakened the fishes' immune systems, so that when the fish encountered harsh winter conditions in 1993, they were unfit and unable to survive. In addition to these problems, the 1989 year class (those herring spawned during the oil spill) failed to recruit into the 1993 herring run, because virtually none of the fish had survived the toxic exposure to oil in their birth year.

This had severe implications for ecosystem recovery, as over forty species of birds, fish, and mammals prey on herring. For example, studies are finding that the lipid-rich herring are the preferred food source for seabird colonies, and that reproductive success correlates directly with lipid content. Without herring, seabirds such as common murres, pigeon guillemots, and black-legged kittiwakes are forced to switch to other available forage fish: capelin, sand lance, pink salmon. But

these fish do not have the high lipid content of herring, and seabird colonies have not recovered to prespill levels in oiled areas. The herring population stabilized in 1995 and 1996, as the weak died and the strong survived. The population still remains well below pre-spill numbers. Harbor seals, killer whales, and sea otters are also listed as not recovering.

Scientists have found that long-term injury is most evident in species such as herring and pink salmon which forage or spend their early life stages in the intertidal zone (the area coated with oil in 1989). Harlequin ducks and sea otters, which both feed heavily on intertidal mussels, are still listed as not recovering. Many intertidal mussel beds and north-facing bays are still heavily contaminated with *Exxon Valdez* oil and pose a serious problem, as scientists are just now realizing. This oil contains polycyclic aromatic hydrocarbons which continue to cause reproductive damage in pink salmon years after the initial spill (Marty et al. 1997).

Meanwhile, from 1989 through 1991, Exxon scientists also conducted studies assessing damage to and recovery of resources injured by its spill. Exxon's studies, which found virtually no long-term effects of oil and a corresponding rapid ecosystem recovery, contrasted starkly with the findings of the *Exxon Valdez* Oil Spill Trustee Council's studies, summarized above. Interested readers are referred to *Sound Truth* (Ott 1994) for a nontechnical comparison of these two data sets to understand the basis for the vast discrepancies in the findings.

2.   The State of Alaska, an ardent road supporter, made steady progress on the project moving east and north across the Delta from Cordova and south from Chitina, despite occasional setbacks from earthquakes and lawsuits. Up until the *Exxon Valdez* oil spill, Cordova residents had spent a good chunk of time each winter debating whether or not the Copper River Highway should be built. More recently, with changing community values, there has been much public dialogue but little true consensus on the issue.

3.   The state and the tourism industry are trying to increase visitor use in the Wrangell–St. Elias National Park and Preserve, in order to relieve visitor impacts in popular Denali National Park to the north. The state intends to pave the existing McCarthy Road from Chitina to its terminus at the Kennicott River, and construction of the much debated Copper River Highway from the interior at Chitina down to the Delta and Cordova would offer visitors and tour operators the option of returning to Anchorage through Cordova and Prince William Sound, rather than along the same route by which they had arrived.

4.   At about the same time, the Minerals Management Service (U.S. Department of Interior) lost interest in promoting an outer-continental-shelf lease sale, which included tracts offshore of the Delta.

Appendix B: The Copper River Watershed Project

1.   The subarctic growing conditions limit productivity of spruce and hemlock on the Delta to 15 to 35 mbf per acre, and saw logs to 7 to 9 mbf per acre. In the Copper Basin, white and black spruce produce 2 to 5 mbf per acre, and saw logs at 1.5 to 2 mbf per acre.

2.   The International Forestry Stewardship Council (IFSC) has offi-

ces throughout the world. In the United States, information about this program can be obtained at IFSC US Initiative, R.D. 1 Box 182, Waterbury, VT 05676. Telephone: 802-244-6257. Fax: 802-244-6258. Email: arp@igc.org.

3.   In 1993, visitation by region of Alaska shows that of the total 836,900 visitors, 68% visited in southcentral, 60% in southeast, 35% in the interior and northern areas, 31% in Denali, and 6% in southwest.

4.   Shorebird Festival and Ice Worm Festival are held on the first weekend in May and February, respectively. Shorebird Festival offers guided birding trips and public transportation to popular staging areas during the day, and evenings of lectures and workshops with international bird experts. Ice Worm Festival (Cordova's solution to mid-winter blues) sponsors two days of competition (such as survival-suit races and the Poker Flats cross-country ski race) spiced with a food-and-crafts fair and a parade. For more information, contact the Cordova Chamber of Commerce, 907-424-7260, or visit the web site http://www.ptialaska.net/ ~ cchamber.

5.   For more information, or to become a member of CRWP, please contact the Copper River Watershed Project, POB 1560, Cordova, AK 99574, telephone 907-424-3334, or email copper@ptialaska.net. The CRWP web site will be available in fall 1998 and will be linked to that of the Cordova Chamber of Commerce.

REFERENCES

Alaska Department of Fish and Game
1994   *Wildlife Notebook Series.* Publication of the Public Communications Section, Juneau.

Allen, Lieutenant Henry T.
1985   *An Expedition to the Copper, Tanana, and Koyukuk Rivers in 1885.* Edited by Terrence Cole. Anchorage: Alaska Northwest Publishing Company.

Arora, David
1986   *Mushrooms Demystified.* Berkeley, California: Ten Speed Press.
        Arvidson, Rose C., Ralph Nichols, and Rose Weathers
1984   *Cordova, the First 75 Years: A Photographic History.* Cordova, Alaska: Fathom Publishing Company; Delta Junction, Alaska: Dragon Press.

Barr, Lou, and Nancy Barr
1983   *Under Alaska Seas. The Shallow Water Marine Invertebrates.* Anchorage: Alaska Northwest Publishing Company.

Berger, Thomas
1985   *Village Journey: The Report of the Alaska Native Review Commission.* New York: Hill & Wang.

Birket-Smith, Kaj, and Frederica de Laguna
1938   *The Eyak Indians of the Copper River Delta, Alaska.* Copenhagen: Det Kal. Danske Videnskab. New York: Selskab.

Blanchet, Dave
1991   Personal communication. Chugach National Forest Hydrological Study. USDA Forest Service, Anchorage.

Burroughs, John, and John Muir
1986   *Alaska, the Harriman Expedition, 1899.* Reprint of 1901 edition. New York: Dover Publications.

Cherrington, Mark
1993   "In the Wake of the Valdez. The Price of Progress Is More Than Just the Cleanup Bill." *Earthwatch* Jan/Feb: 22-29.

Cobb, Clifford, Ted Halstead, and Jonathan Rowe
1995   "If the Economy Is Up, Why Is America Down?" *The Atlantic Monthly.* October.

Cordova Historical Society
1940   "Second Annual Sea Food Festival." Cordova, Alaska. August 10.

Cordova Times
1929   "Send Record Shipment of Canned Crab." April 23.

de Laguna, Frederica
1970   "The Atna of the Copper River, Alaska: The World of Men and Animals." *Folk: Dansk Ethnografisk* 11-12(1969-70):17-26.

Dyer, Christopher, Duane A. Gill, and J. Steven Picou
1992   "Social Disruption and the Valdez Oil Spill: Alaskan Natives in a Natural Resource Community." *Sociological Spectrum* 12:105-26.

Ehrlich, Paul R., David S. Dobkin, and Darryl Wheye
1988   *The Birder's Handbook: A Field Guide to the Natural History of North American Birds.* New York: Simon and Schuster.

Exxon Valdez Oil Spill Trustee Council
1997   *1997 Status Report.* Publication of Alaska Department of Fish and Game, Anchorage.

Fall, James
1990   "Subsistence after the Spill: Uses of Fish and Wildlife in Alaska Native Villages and the *Exxon Valdez* Oil Spill." Publication of Alaska Department of Fish and Game, Anchorage.
1991   "Subsistence Uses of Fish and Wildlife in 15 Alutiq Villages after the *Exxon Valdez* Oil Spill." Publication of Alaska Department of Fish and Game, Anchorage.

Faulkner, Patience Anderson
1992   Unpublished manuscript. Cordova, Alaska.

Fitzhugh, William W., and Aron Crowell
1988   *Crossroads of Continents: Cultures of Siberia and Alaska.* Washington, D.C.: Smithsonian Institution Press.

Ford, Corey
1966   *Where the Sea Breaks Its Back: The Epic Story of a Pioneer Naturalist and the Discovery of Alaska.* Boston: Little Brown and Company.

Grimes, David
1997   "First Do No More Harm." *Alaska's Wild Voices.* (Autumn 1997), Anchorage.

Hanable, William S.
1982   *Alaska's Copper River: The 18th and 19th Centuries.* Alaska Historical Commission Studies in History, no. 21. Alaska Historical Society for the Alaska Historical Commission, Anchorage.

Isleib, M. E. "Pete," and Brina Kessel
1973   *Birds of the North Gulf Coast—Prince William Sound Region, Alaska.* Biological Papers of the University of Alaska, no. 14. Reprinted by University of Alaska Fairbanks.

Janson, Lone E.
1975   *The Copper Spike.* Anchorage: Alaska Northwest Books.

Krauss, Michael E.
1982   *In Honor of Eyak: The Art of Anna Nelson Harry.* Fairbanks: Alaska Native Language Center, University of Alaska.

Kruger, Linda E., and Catherine B. Tyler
1995   *Management Needs Assessment for the Copper River Delta, Alaska.* General Technical Report PNW-GTR-356. United States Department of Agriculture, Forest Service, Pacific Northwest Research Station, Portland, Oregon.

LaDuke, Winona
n.d.   Unpublished. Ponsford, Minnesota.

Lethcoe, Jim
1990   *An Observer's Guide to the Geology of Prince William Sound.* Valdez, Alaska: Prince William Sound Books.

Lethcoe, Nancy
1987   *An Observer's Guide to the Glaciers of Prince William Sound.* Valdez, Alaska: Prince William Sound Books.

Marty, Gary, Jeffrey Short, Donna Dambach, Neil Willits, Ronald Heintz, Stanley Rice, John Stegeman, and David Hinton
1997   "Ascites, Premature Emergence, Increased Gonadal Cell Apoptosis, and Cytochrome P4510A Induction in Pink Salmon Larvae Continuously Exposed to Oil-Contaminated Gravel during Development.' *Can. J. Zool.* 75:989-1007.

Mealy, William V., and Peter Friederici, editors
1992   *Value in American Wildlife Art: Proceedings of the 1992 Forum.* Jamestown, New York: Roger Tory Peterson Institute of Natural History.

Mickelson, Pete
1989   *Natural History of Alaska's Prince William Sound and How to Enjoy It.* Cordova, Alaska: Alaska Wild Wings.

Mitchell, Donald Craig
1997   *Sold American: The Story of Alaska Natives and Their Land, 1867-1959, the Army to Statehood.* Hanover, New Hampshire: University Press of New England.

Nielsen, Nicki J.
1984   *From Fish and Copper: Cordova's Heritage and Buildings.* Alaska Historical Society Commission Studies in History, no. 124. Cordova Historical Society. Anchorage: Thompson Printing.

Ott, Riki, Ph.D.
1994   *Sound Truth: Exxon's Manipulation of Science & the Significance of the Exxon Valdez Oil Spill.* Publication of Greenpeace, Alaska, Anchorage.

Picou, J. S., D. A. Gill, and M. J. Cohen, editors
1997   *The* Exxon Valdez *Disaster: Readings on a Modern Social Problem.* Dubuque, Iowa: Dendall-Hunt Publishers.

Pojar, Jim, and Andy McKinnon, editors
1994   *Plants of the Pacific Northwest Coast: Washington, Oregon, British Columbia, and Alaska.* Alberta, Canada: Lone Pine Publishing.

Power, Michael T.
1996   *Lost Landscapes and Failed Economies: The Search for a Value of Place.* Washington, D.C.: Island Press.

Pratt, Lee C.
1930   *Cordova-The Copper Gateway to Alaska.* In *Cordova Daily Times All-Alaska Review for 1930.*

President's Council on Sustainable Development
1996   *Sustainable America: A New Consensus for Prosperity, Opportunity and a Healthy Environment for the Future.* Washington, D.C.: GPO.

Rennick, Penny, editor
1996   "Native Cultures in Alaska." *Alaska Geographic Society* 23(2):112

Salmon, David K., and Thomas C. Royer
1994   "Aspects of the Meteorology of the Gulf of Alaska." In *Oceanographic and Meteorological Influences on the Upper Gulf of Alaska, Technical Reference Papers Conducted in 1995-1996 for the Area Planning Committee Working Group.* Institute of Marine Science, University of Alaska Fairbanks.

Stegner, Wallace
1982   *Beyond the Hundredth Meridian.* Reprint of 1953 edition. Lincoln: University of Nebraska Press.

Strohmeyer, John
1993   *Extreme Conditions: Big Oil and the Transformation of Alaska.* New York: Simon & Schuster.

Sweetland Smith, Barbara, and Redmond J. Barnett, editors
1990   *Russian America: The Forgotten Frontier.* Tacoma: Washington State Historical Society.

Thomas, Gary, Ed Backus, Harold Christensen, and James Weigand
1991   *Prince William Sound-Copper River-North Gulf of Alaska Ecosystem* Portland, Oregon: James Dobbin Associates and Miles Fridberg Molinari.

Underwood, Paula
1993   *The Walking People: A Native American Oral History.* San Anselmo, California: A Tribe of Two Press.

U S Dept. of Agriculture, Forest Service, Chugach Ranger District
1992   *Take a Hike! Explore the Cordova Ranger District Trail System.* Publication of Chugach Ranger District, Forest Service, Cordova.

Viereck, Leslie A., and Elbert Little, Jr.
1972   *Alaska Trees and Shrubs.* Agriculture Handbook no. 410, Forest Service, U S Dept of Agriculture, Washington, D.C. Reprinted by University of Alaska Fairbanks.

Vitt, Dale H., Janet E. Marsh, and Robin B. Bovey
1988   *Mosses, Lichens, and Ferns of Northwest North America.* Alberta, Canada: Lone Pine Publishing.

Wheeller, Brian
1997   "Tourism's Troubled Times: Responsible Tourism Is Not the Answer." In *Earthscan Reader in Sustainable Tourism*, Lesley France, editor. London: Earthscan Publishers.

Wohlforth, Charles
1998   "Three Gems: Discover a Trio of Towns that Time Forgot — Just Off the Tourist Track." *Alaska Magazine.* February.

ABOUT THE ARTISTS

Abbreviations used: LYWAM for Leigh Yawkey Woodson Art Museum, Wausau, Wisconsin; RSPB for Royal Society for Protection of Birds; SWLA for Society of Wildlife Artists; and SWAN for Society of Wildlife Art for the Nations.

TONY ANGELL (USA)

Born 1940. Graduate of the University of Washington. Primarily a carver of stone, his forms are chosen from the diversity of animal images of the North Pacific. Has written and illustrated a number of award-winning books on nature. Has worked actively as a board member of Washington's chapter of The Nature Conservancy. Participant in the National Academy of Design shows and an elected Fellow of The National Sculpture Society. Shows his work internationally. Lives in Seattle, Washington.

VICTOR BAKHTIN (Russia/USA)

Born 1951. Artist in residence at the International Crane Foundation, Baraboo, Wisconsin, USA, since 1994. Planned to be a violinist, but turned to art after a climbing accident in 1969. Graduated in graphic design from Moscow Printing Institute. Has worked as a hang-gliding instructor and professional illustrator (76 books). Lives in Baraboo, Wisconsin.

DAVID BARKER (New Zealand)

Born 1941. Studied fine art at Auckland University and the University of Hawaii, where he received a Master's degree in Fine Art. Professional painter since 1971; exhibits all over the world. Has taught in Australia, New Zealand, Vanautu, and Canada. Has written and illustrated two books, and has written and directed TV documentaries on underwater life. A keen sailor, he designs and builds ocean-going catamarans. Lives in British Columbia, New Zealand, and England.

DAVID BENNETT (England)

Born 1969. Studied art at Concaster Art College and Leeds Polytechnic. Presently at the Royal College of Art working toward Master's in natural history illustration. In 1993, won first prize in the Swarovski Bird Art Awards and the Sir Peter Scott Memorial Travel Scholarship in the Nature in Art Awards, organized by SWAN. Member SWLA. One-man exhibition at the Wildlife Art Gallery, Lavenham, in 1994. Lives in London.

KEITH BROCKIE (Scotland)

Born 1955. Studied illustration and printmaking at Duncan of Jordanstone College of Art, Dundee. Has traveled from Yemen to the Arctic in search of birds as both bird-bander and artist. Has illustrated numerous bird books. Has written three books on Scottish wildlife and is now working on his fourth. Lives near Invergowrie.

YSBRAND BROUWERS (The Netherlands)

Born 1946. Trained as a garden and landscape designer. Founded Wildlife Art Promotion in 1978 and Artists for Nature Foundation in 1990. Is a keen birder, photographer, and cook who likes to subsist from the land. Lives in Bruinehaar. (Contributing photographer)

VADIM GORBATOV (Russia)

Born 1940. Studied at Academy of Art, Industrial Design and Applied Art, Moscow. Worked as illustrator for Soviet Television, becoming head of graphics and illustration. Has traveled widely throughout former Soviet Union in search of wildlife and has exhibited in The Netherlands, France, UK, and USA. Works as freelance illustrator; has illustrated several books. Lives in Moscow.

DAVID GRIMES (USA)

Born 1952. A musician, storyteller, wilderness guide, and former commercial fisherman, he is co-founder of the Coastal Coalition, a group dedicated to protection and restoration of ecosystems and sustainable economies in the wake of the *Exxon Valdez* oil spill. An adopted member of the Eyak Indian Nation, he lives in Cordova, Alaska, and Santa Cruz, California. (Contributing photographer)

ANDREW HASLEN (England)

Born 1953. Works as a painter and printer of wildlife and domestic animals, art, producing watercolors, oils, sculptures, and linocuts. Influenced by artists Kim Atkinson and Talbot Kelly. Manages Wildlife Art Gallery in Lavenham. Lives in Preston St. Mary, Suffolk.

PAT AND ROSEMARIE KEOUGH (Canada)

Born 1945 (Pat) and 1959 (Rosemarie). Photographers and authors of five best-selling art books, including the acclaimed *The Nahanni Portfolio* (1988). Have directed and appeared in numerous television and video productions featuring nature and wilderness. Won Can Pro Award in 1993 for the best independently produced film. Live on Salt Spring Island, British Columbia.

PIET KLAASSE (The Netherlands)

Born 1918. Studied at Royal Academy of Arts, The Hague. Taught for thirty-four years at Rietveld Academy, Amsterdam, before becoming a full-time artist. Paints landscapes, portraits, and horses. His books include *Jam Session: Portraits of Jazz and Blues Musicians* (1983), and *Concours Hippographique: Illusion of Movement* (1993). Won Golden Bush Award in 1981. Lives in Eemnes.

DYLAN LEWIS (Republic of South Africa)

Born 1964. Studied in Cape Technikson, Cape Town, and the Ruth Prowse School of Art. Worked as a taxidermist in the Rondvlei Nature Reserve, but in 1990 decided to become a full-time painter and sculptor. Built his own studio and foundry, where he produces life-size bronzes of African wildlife. Lives in Cape Town.

PAT MCGUIRE (USA)

Born 1953. Graduate of the University of Washington. Member of the Nature Printer's Society. Has traveled to Japan to study both fish printing and papermaking. Her *gyotaku*, the Japanese art of fish printing and collage techniques, are well known in Alaska, where she has displayed in group, solo, and juried shows and has won numerous grants and awards. Has taught printmaking in the Artists in the School Program. Currently fishes commercially during the summers while spending her winters in the art studio. Lives in Cordova, Alaska.

SUSAN OGLE (USA)

Born 1943. Studied architecture at the University of Cincinnati and painting at Heron School of Art. An award-winning graphic designer in Anchorage for fifteen years before moving to Cordova, Alaska, in 1987 to become a full-time painter. Birding has been a lifelong passion, and the birds of the Copper River Delta and Prince William Sound, a specific study. Has illustrated several children's books relating to the Cordova area.

JOHN PAIGE (England)

Born 1927. Painter, printmaker. Officer in the Royal Amoured Corps for fifteen years, then a game warden in Uganda and a corn merchant before attending Birmingham College of Art and Design. Paints landscapes, portraits, and wildlife, using a variety of techniques, including collage. Member SWLA and Oundle Artists Group, U.K. Presented in Hammond's *Twentieth Century Wildlife Artists* (1986). Work is in various collections, including LYWAM. Lives in Kings Cliffe.

BRUCE PEARSON (England)

Born 1950. Studied Fine Art at the Leicester Polytechnic. Worked for the RSPB Film Unit before turning to art and illustration. Travels widely in search of subject matter, including the Antarctic. Has written and presented a television series on birds and their habitats and has followed birds on migration from the Arctic to Africa. Both themes resulted in *An Artist on Migration* (1991) and *Birdscape* (1991). Lives in Great Gransden.

ANDREA RICH (USA)

Born 1954. Graduate of the University of Wisconsin. Has an international reputation as a printmaker. Works with multiple-color woodcuts. Exhibits extensively and her works are included in many museum and university as well as private collections. Member of the Society of Animal Artists and the California Society of Printmakers. Lives in Santa Cruz, California, with her husband and daughter.

DAVID ROSENTHAL (USA)

Born 1953. After an early interest in physics, became a self-taught landscape painter with an interest in the icy scenes of the northern latitudes and Antarctica. Participated in the National Science Foundation Antarctic Artist and Writer Program as artist-in-residence at McMurdo Station during the austral summer of 1993-94, and during the austral winter 1996. Lives in Cordova, Alaska.

COLIN SEE-PAYNTON (Wales)

Born 1946. Printmaker and painter. Studied at Northhampton College of Art. Started out as a shopkeeper, then worked as a painter. In 1980 produced his first wood engravings, which were used to illustrate *The Book of Jonah* (1987), *Giraldu Cambrensis, Itinerary through Wales* (1989), and others. *The Incisive Eye* (1996) is a catalog raisonné of his wood engravings. Member of the society of wood engravers and the Royal Society of Painters-Printmakers. Work is in collections of the National Museum of Wales; the Ashmolean Museum, Oxford; the Ulster Museum, Belfast; the Berlin Graphotek; and the Museum of Modern Art, Wales. Lives in Powys.

TODD SHERMAN (USA)

Born 1955. Undergraduate degree in Art from the University of Alaska Fairbanks, Master of Fine Art in printmaking from Pratt Institute in Brooklyn, New York. Twenty solo exhibitions and over ninety juried and invitational shows in U.S. Curator and coordinator of twenty exhibitions, notably Relics, Artifacts, and other Mysteries. Assistant professor of art at the University of Alaska Fairbanks. Has big cat artwork in the Nature Company's The Clan of the Wild Cats (1996). Lives in Fairbanks, Alaska, with his wife, Kristi, and three children.

TIM SHIELDS (USA)

Born 1955. Self-taught artist. Master's degree in wildlife ecology from University of California at Riverside. Twenty years of field work studying desert reptiles. Numerous exhibitions in Alaska and the Southwest U.S. Interest in producing images dealing with human effects on ecosystems. Lives in Haines, Alaska.

JUAN VARELA SIMÓ (Spain)

Born 1950. Self-taught artist. Master's degree in animal behavior from Madrid University. Was seabird researcher and director of Spanish Ornithological Society (SEO) 1986-92. Editor of SEO magazine, *La Garcilla*. Author of a game-species identification guide. Illustrator of Joaquin Araujo's *Todavia Vivo* and several other books. Lives at Pozuelo de Alarcon.

SIEGFRIED WOLDHEK (The Netherlands)

Born 1951. Studied biology at the Free University in Amsterdam. Self-taught artist, specializing in cartoons for national Dutch newspapers. Former president of the Dutch Society for Bird Protection, and current director of Worldwide Fund for Nature, Holland. Lives in Soest.

[148] *The artists at Copper River Delta*
In front: D. Barker, J. Varela Simó, R. Keough, P. Keough, S. Ogle
Middle row: T. Angell, A. Rich, V. Gorbatov, K. Brockie, S. Woldhek, A. Haslen
Back row: D. Bennett, D. Rosenthal, D. Lewis, Y. Brouwers

ARTISTS AND SUBJECT INDEX

Page numbers in italics refer to art work.

Ysbrand Brouwers (in Power Creek)
PHOTO: PAT AND ROSEMARIE KEOUGH

Riki Ott
PHOTO: PAT AND ROSEMARIE KEOUGH

In the last decade, I have had the privilege to help organize and guide seven A.N.F. projects, each in very special regions of the world. These projects were concerned with the long-term protection of old cultural landscapes and important fragments of nature including sites in southern and eastern Europe and India. In Alaska, however, on the Copper River Delta, we found an almost untouched ecosystem of vast size, richness, variety, and beauty. For us, this was an unprecedented and unforgettable encounter.

Aldo Leopold once wrote: 'To keep every cog and wheel is the first precaution of intelligent tinkering.' In collaborating with nature, we must not throw away the pieces. If you take a gear or spring out of a clock, the clock stops ticking. The Copper River Delta is just such a clock, and its millions of birds, fish, and mammals, its plantlife, and its people are the vital cogs and springs of a richly interactive ecosystem.

In some of the old cultural landscapes, people have learned, through time and with continuous hard work, to live in grace with their surroundings. The Copper River Delta, too, is dependent for its future upon such respectful and sustainable use of its natural resources.

We invite you to look over the shoulders of the artists who have contributed to this book. We welcome you to share their vision of this magnificent Delta.

Ysbrand Brouwers
Artistic Director of Artists for Nature Foundation

About the Author

A second-generation activist with a doctorate in marine biology from the University of Washington, Riki Ott arrived in Cordova, Alaska, in 1985 intent on spending just the summer. Captivated by the wild country, wild adventures – and equally wild politics – she stayed, fishing a drift gillnet permit for ten years. After the *Exxon Valdez* oil spill in Prince William Sound, she turned more seriously to politics, dedicating her academic training to helping citizens redefine business practices and government accountability to better balance economic, social, and environmental concerns. She has written numerous white papers and received numerous awards for her activism – and for her fiber art quilts, which reflect the beauty of her natural surroundings. She is executive director of the Copper River Watershed Project and chief designer for Hot Yotts, a fiber art company. This is her first book.